condense 凝縮 *gyōshuku*

equivalent 相当する

FLUID AND
PARTICLE MECHANICS

purifier 浄化器 *jōkaki*

evaporator 蒸発器

Filter *firuta* 3過器

mizo & *suiro* 3 channel

filter aid	3過助剤	じょざい
bed	3床	3しょう
cake	3土塊	3かい
cloth	3布	3ぬ
layer	3過層	3かそう
leaf	3葉	3よう
paper	3紙	3し
filter sand	こし砂	こしづな
filtrate	3液	3えき
filtration	3過	

FLUID AND
PARTICLE MECHANICS

BY

S. J. MICHELL, M.SC., D.L.C., C.ENG., M.I.CHEM.E.

Senior Lecturer in Chemical Engineering,
Huddersfield College of Technology

拡 散

単相系の混合物において、その各成分
の間に相対的な移動が行なわれること.

shinto-
滲 透
osmosis
permeation

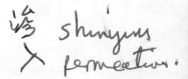

滲
入
shinyun
permeation.

PERGAMON PRESS

OXFORD · LONDON · EDINBURGH · NEW YORK
TORONTO · SYDNEY · PARIS · BRAUNSCHWEIG

Pergamon Press Ltd., Headington Hill Hall, Oxford
4 & 5 Fitzroy Square, London W.1
Pergamon Press (Scotland) Ltd., 2 & 3 Teviot Place, Edinburgh 1
Pergamon Press Inc., Maxwell House, Fairview Park, Elmsford, New York 10523
Pergamon of Canada Ltd., 207 Queen's Quay West, Toronto 1
Pergamon Press (Aust.) Pty. Ltd., 19a Boundary Street, Rushcutters Bay,
N.S.W. 2011, Australia
Pergamon Press S.A.R.L., 24 rue des Écoles, Paris 5ᵉ
Vieweg & Sohn GmbH, Burgplatz 1, Braunschweig

First edition 1970
Library of Congress Catalog Card No. 77–81250

Printed in Hungary

08 013312 6 (flexicover)
08 013313 4 (hard cover)

CONTENTS

PREFACE

THE primary objective of this book is to present an introductory text for undergraduates in chemical engineering, in their early stage of studying the subject. It covers a large part of the syllabuses as set out by the Institution of Chemical Engineers for their external examinations, and to some extent it also meets the needs of the Council of Engineering Institutions (CEI) Part I and Part II examinations. It is therefore hoped that the book will also be of some interest to students undergoing courses other than chemical engineering.

The first seven chapters of the book are devoted to what is commonly regarded as fluid mechanics, or hydraulics, while the last two chapters deal almost exclusively with that part of the subject which is traditionally thought to belong to chemical engineering. The main emphasis of the book is therefore understandably on physical concepts rather than on highly mathematical treatment.

Each chapter of the book is illustrated by a number of numerical examples in support of the theory, and problems for the student to test his grasp of the principles will be found at the end of each chapter. A selected number of references has also been appended for those who may wish to expand their knowledge of the subject beyond the limitations of this text.

The text is heavily biased towards the British system of units, with the gravitational conversion factor (g_c) preserved in conformity with existing engineering practice. Should, however, need arise to make use of the underlying theory in an absolute system of units, such as SI, for example, this conversion factor can be simply removed from the equations considered without impairing their validity.

The author is deeply grateful to his wife Jadwiga for her encouragement and invaluable help given in the preparation of the manuscript, and to his son Richard for his competent proof-reading.

S. J. MICHELL

CHAPTER 1

GENERAL PRINCIPLES

T:IE word mechanics implies some connection with machinery. In the field of science, however, this word is more closely associated with the study of the behaviour of objects under the action of forces. In fluid and particle mechanics, the objects are small particles of gas, liquid or solid which form the bulk of matter under study.

Fluid mechanics is concerned with the properties and behaviour of liquids and gases in motion and at rest. The science of fluid mechanics is accordingly subdivided into two main branches, namely fluid dynamics and fluid statics. The former deals with the relation between the behaviour of fluids in motion and the forces which produce that motion, while statics may be regarded as a special case of dynamics when the net effect of forces acting on the fluid body produces no motion relative to a boundary.

A branch of dynamics which deals with motion from the geometrical point of view, but without reference to the forces causing the motion, is called kinematics. Hydraulics, hydrodynamics and aerodynamics, as well as particle mechanics, are specialised branches of fluid mechanics.

1.1. Historical Outline

The Egyptians are known to have had some knowledge of hydro-dynamics as early as 3000 B.C. in connection with shipbuilding. Such knowledge as existed then was, however, gradually lost in the effort to construct ships of larger size. Centuries passed and, with the exception of the discoveries of Archimedes (287–212 B.C.) in fluid statics, there is no record of any progress in fluid mechanics until the time of Leonardo da Vinci (1452–1519). One of the most distinguished masters of art of the

1

Renaissance, Leonardo is also noted for his keen interest in the field of science and engineering. The full account of his pioneering work in this field will probably never be available, as—apart from his famous treatise on painting—no attempts were made early enough to publish his manuscripts. But what remains indisputable is that Leonardo recovered Archimedes' knowledge of the pressure of liquids, and that in his innumerable researches into the field of hydro- and aerodynamics he was far ahead of his time.

Towards the end of the sixteenth century Galileo (1564–1642) emerged as a brilliant physicist and experimentalist. His famous experiments on falling bodies, as well as the less famous but equally important experiments on rolling balls down inclines and his investigations on the motion of projectiles, paved the way for the outstanding achievements of later generations. The law of motion, which Galileo anticipated in his treatment of the horizontal component of the motion of projectiles, was later incorporated by Newton (1642–1727) into a comprehensive set of laws of motion.

The science of fluid dynamics may be said to have originated with the experiments of Torricelli (1608–1647) on the discharge of water from orifices. His main contribution to the science lies in associating the velocity of efflux of a liquid from an orifice with the velocity of falling bodies. Torricelli also gave the first description of a barometer (1644), hence the name "Torricellian vacuum" for the space above the mercury in a sealed tube. His French contemporary, Pascal (1623–62), supplemented the work of Torricelli on barometers and particularly on a vacuum, and his treatise on the equilibrium of liquids (1663) opened the way for the hydraulic press. Pascal is also supposed to have performed experiments on the compressibility of air but, as no records of this work exist, the distinction goes to Boyle (1627–91) and Mariotte (1620–84). The latter had also turned his attention to flow in canals and pipes, and was probably the first to observe that the resistance to flow is proportional to the square of the velocity, which—as we now know—is about correct for turbulent flow only. Newton seems to have intercepted the work of Mariotte while formulating his basic resistance equation from consideration of the change of momentum of fluid around a moving object.

In 1738 Daniel Bernoulli (1700–82) published a classical work, famous for the theorem which bears his name. Bernoulli's guiding principle was that of the conservation of energy, and the theorem, which appears in various forms in all classical treaties on hydrodynamics, marks the advent of the modern science of mechanics of fluids. In 1750 Euler (1707–83), arrived at a mathematical formulation similar to that developed by Bernoulli, by the application of the momentum principle to inviscid flow. The theory has subsequently been greatly expanded by the contribution of a number of mathematicians, the most notable being D'Alembert (1717–83), Lagrange (1736–1813), Laplace (1749–1827), Navier (1785–1836), de Cauchy (1789–1857), and Stokes (1819–1903).

The purely theoretical approach to the study of fluid mechanics, based on the simplifying assumption that fluids are ideal, could not satisfy engineers. Faced with practical problems, they had to diverge from the school of thought led by the mathematicians, and to establish a new one based largely—if not entirely—on empiricism. Both schools have persisted down to the present day, each contributing magnificently to the subject. The divergence of points of view between the two schools has, however, never been complete. More often than not, the members of apparently one school have turned their attention to the point of view of the other school, thus bringing about the reconciliation between the two. Reynolds (1842–1912), for instance, though an engineer by training and by his early experiments, achieved considerable repute in the field of theoretical science. Similarly, Navier, already classed as a mathematician for his contributions to hydrodynamics, was also an engineer by both education and vocation. The two examples illustrate how difficult it is to allocate a name to one school or another.

The present trend is generally consistent with earlier developments in both theoretical and applied science. The beginning of this century is, however, interesting for a new approach to the analysis of fluid motion. The distinction goes to Ludwig Prandtl (1875–1953), a mechanical engineer by upbringing. His concept of the boundary layer has had a great influence upon the line of research in hydraulics, aeronautics and other related fields. The work of Prandtl was supplemented to a high degree by von Karman (1881–1963), who provided an analytical background to the rather physical approach of Prandtl.

Apart from the above, many names would be mentioned but for lack of space. These, however, like many other names which have made the history of the subject of fluid mechanics, can be traced in the most interesting historical survey of Rouse and Ince.[1]

1.2. Kinematics

Kinematics is the study of the geometry of motion without reference to the forces causing it. In the subject of fluid mechanics, kinematics assumes that fluid is made up of a large number of very small elements, called particles, in close contact. These are considered as geometrical points of a definite mass and volume.

Consider fluid particles moving in succession along a curved line in Fig. 1.1. At any instant, the particles will have a definite velocity at any point on the curve. There are two important names attached to such a situation. The motion is said to be *steady* when the particles arriving at

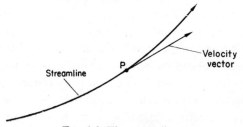

FIG. 1.1. The streamline.

a given point, such as the point P on the curve, in Fig. 1.1, have the same velocity at all instances, though they may have different velocities at different points on the curve. If, in addition, the particles have also the same velocity at all the points forming the curve, at any instant considered, the motion is said to be *uniform*. It follows that a uniform motion may also be a steady one.

The actual motion of fluid particles in a confined space is rather complex. For the sake of simplicity, however, particles are assumed to follow a definite path, and the line formed by a succession of such particles is called the *streamline*.

A streamline is defined as a line drawn tangent to the velocity vectors of particles forming the line. This definition ignores the fact that particles may actually be moving in an erratic motion instead of following strictly continuous lines, hence the velocity vectors which define streamlines must be taken as representing the net effect of the motion. The slower the motion the more exact picture is reproduced by the streamline. Slow motion is therefore called *streamline* or *laminar* motion. In motion of this kind, viscosity of the fluid in motion is more important in shaping the stream-lines than the velocity itself and for this reason a streamline motion is also called *viscous* motion. The reverse is true in fast motion when fluid particles frequently cross and recross the streamlines. Such a motion is called *turbulent*.

A bundle of streamlines forms a tubular surface which is called the *streamtube* (Fig. 1.2). First introduced to the subject of fluid mechanics by Borda (1733–99) in connection with his investigation of discharge from tanks, the concept of streamtube has been found very useful in analysing problems of an apparently different nature. For example, such problems

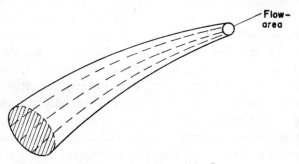

FIG. 1.2. The streamtube and the flow area.

as flow in conduits and flow around solid objects may be treated by the application of the same laws. In each case the problem is approached by consideration of a streamtube in isolation from the adjacent fluid.

A streamtube may be cut by any number of surfaces drawn at right angles to the tube. Such a surface is called the *flow-area*. Using the terminology adopted for single streamlines, the flow is said to be *steady* if at any point of flow-area the net velocity vector remains unchanged with time. Conversely, the flow is *unsteady* if the net velocity vector at the point considered changes with time. This kind of flow will not be considered in this text. Also, the flow is said to be *uniform* if the net velocity vectors at all the points of the flow-area have the same magnitude. Such an idealized flow is often assumed in theoretical considerations.

1.3. Units and Dimensions

A standard quantity with which others of the same kind are compared for purposes of measurement is called a unit.

Only three independent units are required to describe a physical quantity in the absence of heat effects. The three units are called the primary, absolute or fundamental units. They can be used to built up any coherent system of subsidiary units, known as the secondary or derived units. If heat effects are to be accounted for, one more fundamental unit is required, and this is usually the unit of temperature.

The units of length, mass and time have been arbitrarily chosen as the three fundamental units in mechanics. The fact that, in some specialised branches of mechanics, force is used instead of mass as one of the fundamental units will be ignored in the text.

There are two sets of fundamental units in common use, and these give rise to two basically different systems of units. The metric system has the centimetre (cm), gram (gm), and second (sec) as the units of length, mass and time, respectively. From their initials the system is known as the cgs system. The British system, which is also widely used in the United States, is based on the corresponding units of foot (ft), pound (lb) and second, and is referred to as the fps system. Both systems are often used interchangeably, the cgs system being almost in universal use in the scientific field, while in engineering work the fps system is preferred. Another metric system for use in both scientific and engineering work is based upon the metre as the unit of length, the kilogram as the unit of mass, and the second. It is known as the mks system.

A refinement of the metric system leads to the so-called SI system (Système International d'Unités). The system has recently been adopted by the International Organisation for Standardisation (ISO), and Great Britain (along with many other countries) has declared its general policy of going over to SI as soon as practically possible.

Six basic primary and a number of secondary (derived) units emerge from the SI system, with several new names attached to some of them. The four primary units which are of interest in the subject of Fluid and Particle Mechanics are the metre (m), kilogram (kg), second (s), and degree Kelvin (°K), the bracketed terms indicating the symbols attached to the corresponding units. The system permits the use of the degree Celsius (°C) along with the degree Kelvin, and accepts multiples and fractions, but in thousandths only, with probably one or two exceptions. For example, the acceptable units of length are metres, kilometres (km), and millimetres (mm), but the handy centimetre has been excluded from this system. Its fate is also shared by a number of such long-standing names as the calorie and horsepower. A table of these units and their conversion factors is appended (see page 334).

The important secondary units in the subject of mechanics are those of velocity, acceleration and force. The first two are obtained by dividing a unit of length by the unit of time or by its square, respectively. The conversion of these units is simple, unlike that of units of force which will be discussed in some detail in the next paragraph.

Dimension is a term used in geometry to denote magnitude in a specified direction. In mechanics, the term is also used with some special meaning with reference to the units adopted in this subject.

Each unit may be represented by a letter which symbolises that unit generally. This letter is then called the dimension of the unit it represents. The letters which have been adopted in this text for the four fundamental units are: L (for length), M (for mass), θ (for time), and T (for temperature). Following the definitions, the dimensions of such secondary units as velocity and acceleration emerge as L/θ and L/θ^2, and that of force as ML/θ^2.

1.4. Force, Mass and Weight

Anything that changes, or tends to change, the state of rest or motion in a body is called force.

Let the symbols F, m, and a represent the force, mass, and acceleration, respectively; then from Newton's second law of motion

$$F = kma \qquad (1.1)$$

where k is a proportionality constant.

The equation can be solved for any set of units, provided that a proper value is assigned to the proportionality constant.

By making the constant dimensionless, and giving it the numerical value of unity, the force will be expressed in the absolute system of units. Thus, depending on the units chosen, two metric and one British units of force may be set up.

In the cgs system, the absolute unit of force is called the *dyne*. It may be defined from eq. (1.1) as the force that gives to a mass of 1 gm an acceleration of 1 cm/sec^2. Similarly, a force which will give an acceleration of 1 m/sec to a mass of 1 kg will define the unit in the mks system. This unit is called the *Newton*. Finally, the force that gives an acceleration of 1 ft/sec to a mass of 1 lb defines the absolute unit of force in the fps system. The unit is called the poundal.

In another system, the constant k of eq. (1.1) is made a dimensional quantity. It has been given a special designation, such that $k = 1/g_c$, where the new constant g_c has the numerical value assigned to the gravitational acceleration at sea level at 45° latitude. As the gravitational acceleration determines the pull of the earth upon a body, a system based on this constant is known as the *gravitational system* of units. The system is widely used in engineering practice, and for this reason it is also called the *engineering system* of units. For this system, eq. (1.1) is more conveniently presented in the form

$$Fg_c = ma \tag{1.2}$$

In the British system of units, the gravitational unit of force has the same name as the unit of mass, namely the *pound*. To avoid ambiguity, the unit of force is called the *pound-force*. In this text, this unit will be described by the symbol *lb-f*, while the symbol for the unit of mass will be left as *lb* without qualification. On this basis, the gravitation constant will be defined by eq. (1.2) as

$$g_c = 32 \cdot 2 \frac{\text{(lb)}}{\text{(lb-f)}} \times \frac{\text{(ft)}}{\text{(sec}^2)} \tag{1.3}$$

in which 32·2 is the numerical value of the standard gravitational acceleration, approximated to three significant figures. The corresponding figure in the cgs system is 981.

A mass which is given an acceleration of 1 ft/sec² by a force of one pound-force is sometimes used as a unit of mass. This unit is called the *slug*. Its use seems to be more confusing than helpful and for this reason the unit will not be considered in this text.

The conversion of the engineering to the absolute units of force, and vice versa, is made by means of the gravitational constant g_c, as defined by eq. (1.3). If this constant is removed from eq. (1.2), by making it unity, the resulting equation is eq. (1.1). It follows that an engineering unit of force is g_c times bigger than the absolute unit. Then

$$1 \text{ lb-f} = 32 \cdot 2 \text{ poundals}$$

The distinction between mass and weight is often the source of misinterpretation because of the subtlety of the concept on which it is based. The concept is simply contained in the statement that mass is the inertia which an object exhibits in response to any effort to start or stop or to change its state of motion, whereas weight is the force of gravitational attraction upon the object.

The statement may be expanded by considering a body falling freely. If *g* is the measured acceleration under the action of the force of gravitational attraction then, from eq. (1.2),

$$Fg_c = mg \qquad (1.4)$$

In this equation, the force F represents a quantity which is commonly known as weight. Since g is a variable and g_c is a true constant, it follows that weight, unlike mass, varies with location. The variation, however, is small enough to be ignored in engineering problems.

1.5. Density

A useful combination of the units of mass and length results in a characteristic known as mass density, or simply *density*. This characteristic property is defined as the amount of mass of a given matter contained in a unit volume. If m and V symbolise mass and volume respectively, then the algebraical statement of the definition is

$$\varrho = \frac{m}{V} \qquad (1.5)$$

where the Greek letter ϱ is the symbol usually given to the density.

The unit of density depends on the units in which the mass and volume are specified. It is common practice to specify them either in the cgs or fps system of units. Taking water as an example, its density at 15·5°C is 1 gm/cm³, or

$$\frac{1/454}{1/(30 \cdot 5)^3} = 62 \cdot 4 \text{ lb/cft.}$$

Unless the temperature warrants some corrective measure, the two figures are commonly used for the density of water in problems referring to the subject of fluid mechanics.

The concept of weight, discussed already, has the same effect on density as it has on mass. The characteristic quantity based on this concept is often called the *weight density*. If the Greek letter γ be the symbol given to this quantity, then following the definition of weight, as given by eq. (1.4), the two densities are related to each other by

$$\gamma = \varrho \left(\frac{g}{g_c} \right) \qquad (1.6)$$

Extensive use is made of this equation in engineering practice.

It may be noted that the numerical values of the gravitational constant g_c and of the variable gravitational acceleration g are identical when approximated to three significant figures. The ratio of the two may therefore be taken as unity. Following the argument, it is evident that—although the two symbols define the density in different dimensions—they also express it at the same numerical level. Taking water again as an example, its weight density is

$$\gamma = 62 \cdot 4 \frac{\text{lb}}{\text{ft}^3} \left[\frac{\text{ft/sec}^2}{(\text{lb/lb-f}) \, (\text{ft/sec}^2)} \right]$$

which simplifies to

$$\gamma = 62 \cdot 4 \frac{\text{lb-f}}{\text{ft}^3} = 62 \cdot 4 \frac{\text{lb-wt}}{\text{ft}^3}$$

Several methods are available for the determination of the density of fluids. For most liquids, this property can be obtained from the relevant literature. For gases, however, this information is less readily available, but also unnecessary. The common procedure leading to the evaluation of the density of a gas is based on the assumption that all gases behave ideally, i.e. that they obey the ideal gas law which is defined by the so-called equation of state

$$pV_m = RT \tag{1.7}$$

Vapours, like steam, are also assumed to obey the law.

The symbols in terms of which the equation is stated represent p = pressure, T = absolute temperature, V_m = specific volume, mole basis, R = universal gas constant.

For one set of units, the constant R has a fixed numerical value. It can be obtained from eq. (1.7) by substituting for the variables their numerical values, which they possess at the so-called standard condition. Normal barometric pressure and the freezing-point of water at this pressure represent the standard pressure and temperature, respectively. The standard specific volume is defined as the volume occupied by one mole of an ideal gas at the standard temperature and pressure. The abbreviations STP (standard temperature and pressure) and SCF (standard cubic feet) are frequently used to describe this condition. At the STP, the specific volume of an ideal gas has a value of 22·4 litres per gram-mole, or 359 ft³ per pound-mole, both approximated to three significant figures.

Let

$$V = \frac{V_m}{M_w}$$

where M_w is the molecular weight of the gas, then eq. (1.7) becomes

$$pV = \frac{RT}{M_w} \qquad (1.8)$$

in which V represents the volume occupied by a unit mass. Its reciprocal is the density, so that $V = 1/\varrho$, and

$$\frac{p}{\varrho} = \frac{RT}{M_w} \qquad (1.9)$$

This equation is quite satisfactory for the determination of the density of gases at conditions normally encountered in fluid mechanics. At higher temperatures and pressures, however, some modification of this equation may be found necessary. The same applies to gases in flow at velocities close to the velocity of sound.

The *specific gravity* of a substance is the ratio of its density and the density of some reference substance. Water at 15·5°C is the reference substance for liquids and solids, and air is often taken as the reference for gases.

EXAMPLE 1.1

What is the density of air ($M_w = 29$) at a temperature of 60°F, and an absolute pressure of 180 lb-f/in²?

Solution

In the fps system of units, the standard pressure is (14·7)(144) lb-f/ft², and the standard temperature is $460 + 32 = 492$°F (abs.). From eq. (1.7)

$$R = \frac{pV_m}{T} = \frac{(14 \cdot 7)\,(144)\,(359)}{492} \quad \frac{(\text{lb-f/ft}^2)\,(\text{ft}^3/\text{lb-mole})}{°\text{F}}$$

Simplifying

$$R = 1544 \frac{(\text{lb-f})\,(\text{ft})}{(\text{lb-mole})\,(°\text{F})}$$

Substituting this value and the data in eq. (1.9) and solving for the density

$$\varrho = \frac{pM_w}{RT} = \frac{(180)\,(144)\,(29)}{(1544)\,(460 + 60)}$$

$$\varrho = 0 \cdot 94 \text{ lb/ft}^3$$

1.6. Pressure and Pressure Head

The force exerted upon a surface of unit area is called the intensity of pressure, or just *pressure*. Like force itself, pressure may be quoted in fundamental or engineering units. The commonly adopted unit of pressure in engineering work is the pound-force per square foot (lb-f/ft^2), or per square inch (lb-f/in^2), usually abbreviated to psf and psi, respectively.

The ratio of the pressure and the density of the fluid to which the pressure refers, when expressed in units of length, is called the *pressure head*. Let p be the pressure, expressed in lb-f/ft^2, in a fluid of weight density γ, then the ratio p/γ has the dimension

$$\frac{\text{lb-f/ft}^2}{\text{lb-f/ft}^3} = \text{ft}$$

Similarly, it can be proved that, in the absolute system of units, the ratio $p/\varrho g$ also has the dimension of length. Thus, taking $p = F/A$, the dimension of pressure emerges as

$$\frac{ML/\theta^2}{L^2} = \frac{M}{L\theta^2}$$

and since the dimensions of density and gravitational acceleration are M/L^3, and L/θ^2, respectively, then the ratio has the dimension of

$$\frac{M/L\theta^2}{(M/L^3)(L/\theta^2)} = L$$

The concept of pressure head explains the variety of units in which pressure is often expressed. As an example, the standard pressure of 1 atmosphere is equivalent to a pressure head of 760 mm of mercury, which corresponds approximately to 760/25·4 = 29·9 in. of mercury, or to (760/25·4) (13·6/12) = 33·9 ft of water, often approximated to 34 ft. Also, since a column of water 33·9 ft high and 1 ft^2 in cross-section weighs 33·9×62·4 = 2115 lb, then this figure also represents the atmospheric pressure in lb-f/ft^2, which is equivalent to 2115/144 = 14·69 lb-f/in^2. This figure is commonly approximated to 14·7 psi.

The pressure in excess of atmospheric pressure is called *gauge pressure*. To distinguish it from absolute pressure, i.e. the pressure quoted above a vacuum, the gauge pressure is abreviated to psfg or psig. Also, to avoid ambiguity, the abbreviation of absolute pressure is often expanded to psfa, or psia.

Subatmospheric pressures are normally reported in terms of the so-called *vacuum*. The conversion from vacuum to absolute units of pressure involves subtraction of the vacuum from the barometric pressure. For instance, 6 in. of vacuum (which means 6 in. of mercury head below atmospheric pressure), at 750 mm barometric reading, corresponds to the absolute pressure of

$$750 - \left(\frac{6}{29 \cdot 9}\right)(760) = 598 \text{ mm mercury,} \quad \text{or}$$

$$598\left(\frac{33 \cdot 9}{760}\right) = 26 \cdot 6 \text{ ft water,} \quad \text{or}$$

$$26 \cdot 6\left(\frac{62 \cdot 4}{144}\right) = 11 \cdot 5 \text{ psia.}$$

1.7. Classification of Fluids

The term fluid may be defined as a substance which undergoes continuous deformation when subjected to a shearing force. This force will be later found to be closely related to an important property of fluids, namely their viscosity. Together with another important property, density, viscosity is used in the classification of fluids from a number of standpoints.

One classification divides fluids into two broad groups, namely *compressible* and *incompressible* fluids, according to the susceptibility of their density to the changes in pressure. The density of liquids is practically independent of pressure, unlike that of the gases. On the basis of this distinction, liquids are included in the class of incompressible fluids, while gases are generally classed as compressible. The demarcation line drawn in this way is, however, far from complete. A gas is often treated as incompressible at least as far as calculations are concerned, if the variation in pressure and temperature within a system considered does not affect its density appreciably.

Another classification divides fluids into *ideal* and *non-ideal*, or real fluids. An ideal, or *perfect*, fluid is a hypothetical liquid or gas which is incompressible and has zero viscosity. All real fluids have finite viscosity, as evident from frictional phenomena in flow, and for this reason they may also be called *viscous* or *viscid* fluids. It follows that the term *inviscid* may be applied to a fluid having zero viscosity. The term is actually often used to distinguish an inviscid fluid from an ideal fluid, which is assumed to be incompressible in addition to being inviscid. An ideal, or perfect, fluid is therefore also inviscid. Although all fluids found in nature belong to the class of real fluids, it is often convenient to class them as ideal, or at least as inviscid. This simplified approach is most commonly met with in classical hydrodynamics.

Finally, fluids may be said to belong to either the *Newtonian* or *non-Newtonian* class. The basis of this classification will be discussed in the next paragraph.

1.8. Viscosity

The frictional phenomena exhibited by real fluids in flow are caused by their most intrinsic property known as *viscosity*. This property can therefore be determined in terms of the force required to overcome the resistance experienced in flow. A simple approach to this problem is based

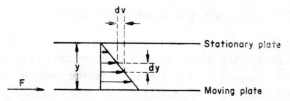

Fig. 1.3. Velocity profile in Couette type flow.

on the consideration of laminar flow between parallel planes when one is stationary while the other is assumed to move in its own plane. The idea was introduced to this subject by Couette in 1890, and flow of this kind is sometimes referred to as *Couette flow*.

Consider a fluid contained between two parallel plates, a short distance apart, as shown in Fig. 1.3. Let the lower plate be set in slow motion parallel to the upper plate which remains stationary. On the surface of each plate there will be a thin layer of the fluid that adheres to it. If we now imagine the rest of the fluid to be made up of a large number of such thin layers parallel to the plates, these layers will be found moving at different velocities, and the resulting velocity profile will be represented by a triangle. This indicates that in the Couette type of flow the velocity gradient, dv/dy, is constant. If this is in agreement with the observed fact, then the force required to produce the gradient must be proportional to it. Let this force be F; then

$$F \propto dv/dy$$

Under this force fluid undergoes continuous deformation, in the same way as an elastic material subjected to a shearing force within the limit of its elasticity. This force is therefore a shearing force for the fluid. Let now

$$\tau = \frac{F}{S} \tag{1.10}$$

where S is the surface area of the moving plate. The Greek letter τ (tau) then represents the shearing force per unit area, i.e. the *shear stress*. The proportionality may then be presented in the form

$$\tau = -\mu \left(\frac{dv}{dy} \right) \tag{1.11}$$

where the other Greek letter μ (mu) is a proportionaiity constant. The constant is called the *coefficient of viscosity*, or simply *viscosity*, and the negative sign which appears in the equation implies that there is a decrease in velocity in the direction of the stationary plate. Equation (1.11) is also a mathematical formulation of a statement known as Newton's law of viscosity. In consequence, fluids that obey the law are called *Newtonian fluids*, the term *non-Newtonian* being attached to very viscous liquids, pastes and plastics which do not obey the law. All fluids considered in this text will belong to the class of Newtonian fluids.

The symbol τ, in eq.(1.11), represents a force per unit area, then its dimension is

$$\frac{ML/\theta^2}{L^2} = \frac{M}{L\theta^2},$$

and since the dimension of the velocity gradient is $1/\theta$, it follows that viscosity has the dimension

$$\frac{M}{L\theta^2} \div \frac{1}{\theta} = \frac{M}{L\theta}$$

In the cgs system, the unit of viscosity is called the *poise*. It has the dimension of (gm/cm) (sec). Since

$$1 \text{ gm} = \frac{(1 \text{ dyne}) (1 \text{ sec}^2)}{(1 \text{ cm})}$$

then the unit may also be expressed in terms of force as

$$1 \text{ poise} = \frac{(1 \text{ dyne}) (1 \text{ sec})}{(1 \text{ cm}^2)}$$

This is equivalent to

$$\frac{(1 \text{ dyne})(1 \text{ cm})}{(1 \text{ cm}^2)(1 \text{ cm/sec})}$$

from which the poise may be defined as the viscosity that requires a shear stress of 1 dyne/cm² to maintain a relative velocity of 1 cm/sec between two planes 1 cm apart.

For most fluids, the unit of poise is inconveniently large, and for this reason viscosity data are usually reported in units 100 times smaller. The unit is called the *centipoise*, abbreviated to cp, so that

$$1 \text{ cp} = 0.01 \text{ poise}$$

The two counterparts in the fps system are

$$\frac{\text{lb}}{(\text{ft})(\text{sec})} \quad \text{and} \quad \frac{(\text{poundal})(\text{sec})}{\text{ft}^2}$$

The conversion factor to bring poises to the corresponding absolute units of the fps system is obtained by taking 453·6 g for 1 lb, and 30·48 cm for 1 ft, hence

$$1 \text{ poise} = \frac{1/453 \cdot 6 \text{ (lb)}}{1/30 \cdot 48 \text{ (ft)} \times 1 \text{ sec}}$$

$$= 0.0672 \frac{\text{lb}}{(\text{ft})(\text{sec})}$$

$$= 0.0672 \frac{(\text{poundal})(\text{sec})}{\text{ft}^2}$$

or, on an hourly basis

$$1 \text{ poise} = 0.0672 \times 3600$$

$$= 242 \frac{\text{lb}}{(\text{ft})(\text{hr})}$$

The more widely used conversion factor is

$$1 \text{ cp} = 2 \cdot 42 \frac{\text{lb}}{(\text{ft})(\text{hr})}$$

$$= 2 \cdot 42 \frac{(\text{poundal})(\text{hr})}{\text{ft}^2}$$

The conversion to engineering units is made by taking

$$1 \text{ lb-f} = 32 \cdot 2 \text{ poundals}$$

The term *relative viscosity* is used occasionally to describe how many times the viscosity of a given fluid is larger or smaller than the viscosity of water at a reference temperature of 68°F. At this temperature water has a viscosity of 1 cp exactly. It follows that in the cgs system relative and absolute viscosities have the same numerical value.

In many equations viscosity and density appear together in the form of their ratio. It is therefore often convenient to use this ratio rather that the two quantities separately. The quantity is known as the *kinematic viscosity*, and the usual symbol assigned to it is the Greek letter ν (nu), so that

$$\nu = \frac{\mu}{\varrho} \tag{1.12}$$

In the cgs system, the unit is called the *stokes*. It is expressed in cm^2/sec. The corresponding unit in the fps system is expressed in ft^2/sec, and has no name. Kinematic viscosities may be quoted in *centistokes*.

Several methods are available for the determination of the viscosity of fluids. Some will be referred to in the latter chapters, but one method which makes direct use of eq. (1.11) will be mentioned at this stage. It is based on rotating a cylinder inside another stationary one, with the annular space being filled with a test fluid. The viscosity is calculated from the moment needed to be applied to hold the outer cylinder stationary while the inner rotates. The application of this method is illustrated by one of the following numerical examples.

The viscosity of liquid decreases appreciably with temperature, while for gases the effect of temperature is less marked. Generally, however, there is a slight increase of viscosity of gases with the rise in temperature. Pressure has practically no effect on the viscosity of both liquids and gases.

EXAMPLE 1.2

The viscosity and density of an oil are 9 poises and 0·9 gm/cm³, respectively. What is the kinematic viscosity of the oil?

Solution

In the cgs system, from eq. (1.12)

$$\nu = \frac{\mu}{\varrho} = \frac{9}{0·9}$$

$$\nu = 10 \ cm^2/sec$$

In the fps system

$$\mu = (9)\,(0{\cdot}0672)\ \text{lb/(ft)}\,(\text{sec})$$

$$\varrho = (0{\cdot}9)\,(62{\cdot}4)\ \text{lb/cft}$$

$$\nu = \frac{(9)\,(0{\cdot}0672)}{(0{\cdot}9)\,(62{\cdot}4)}$$

$$\nu = 0{\cdot}0108\ \text{ft}^2/\text{sec}$$

EXAMPLE 1.3

A liquid fills the space between two parallel plates 0·1 ft apart. What is the viscosity of the liquid if a force of 27·64 dynes is required to move one plate with a uniform velocity of 0·1 ft/sec while the other plate remains stationary? The moving plate has an area of 1 ft².

Solution

Referring to Fig. 1.3

$$-\frac{dv}{dy} = \frac{v}{y} = \frac{0{\cdot}1}{0{\cdot}1} = 1{\cdot}0\ \text{sec}^{-1}$$

where v is the velocity of the moving plate.

$$1\ \text{poundal} = (1\ \text{lb})\,(1\ \text{ft/sec}^2) = (453{\cdot}6\ \text{gm})\,(30{\cdot}48\ \text{cm/sec}^2)$$
$$= 13{,}820\ \text{dynes}$$

Then the force required to keep the plate moving at this velocity is

$$F = \frac{27{\cdot}64}{13{,}820} = 0{\cdot}002\ \text{poundals}$$

From eq. (1.10)

$$\tau = F/S = 0{\cdot}002/1{\cdot}0 = 0{\cdot}002\ \text{poundals/ft}^2$$

From eq. (1.11)

$$\mu = \frac{\tau}{-(dv/dy)} = \frac{0{\cdot}002\ \text{poundal/ft}^2}{1{\cdot}0\ \text{sec}^{-1}}$$

$$\mu = 0{\cdot}002\ \frac{\text{poundal} \times \text{sec}}{\text{ft}^2}$$

or

$$\frac{0{\cdot}002}{0{\cdot}0672} = 0{\cdot}02976, \quad \text{say} \quad 0{\cdot}0298\ \text{poise}, \quad \text{or} \quad 2{\cdot}98\ \text{cp}$$

EXAMPLE 1.4

A cylinder of 0·4 ft radius rotates coaxially inside a stationary cylinder of 0·5 ft radius. Both cylinders are 2 ft long. What is the viscosity of the liquid which fills the annular space if a torque of 0·463 (ft) (lb-f) is required to maintain a speed of rotation of 60 rpm?

Solution

Torque applied = Resisting torque = Resisting force × arm
The resisting torque = $(F)(r) = (0\cdot463)(32\cdot2) = 14\cdot91$ (poundal) (ft)
From eq. (1.10)
$$F = (S)(\tau) = (2\pi rL)(\tau)$$
Since $L = 2$ ft
$$F = 4\pi r\tau$$
where r is the variable radius.
It follows that
$$(F)(r) = 4\pi r^2\tau = 14\cdot91 \text{ (poundal) (ft)}$$
from which
$$\tau = \frac{14\cdot91}{4\pi r^2} = \frac{1\cdot187}{r^2}, \quad \text{poundal/ft}^2$$

The tangential velocity at the stationary cylinder is $v_o = 0$, and that at the rotating inner cylinder is
$$v_i = \omega r_i = \frac{2\pi N r_i}{60}$$

Since $N = 60$ rpm, and $r_i = 0\cdot40$ ft, then $v_i = 2\cdot512$ ft/sec. Taking distances between the cylindrical liquid layers in terms of their radii inwardly $-dy = dr$, and from eq. (1.11)
$$(dv) = \frac{\tau(dr)}{\mu} = \left(\frac{1\cdot187}{r^2}\right)\left(\frac{dr}{\mu}\right)$$

Integrating
$$\int_{v_o}^{v_i} dv = \frac{1\cdot187}{\mu} \int_{r_o}^{r_i} \frac{dr}{r^2}$$
$$v_i - v_o = \frac{1\cdot187}{\mu}\left(\frac{1}{r_i} - \frac{1}{r_o}\right)$$
$$2\cdot512 - 0 = \frac{1\cdot187}{\mu}\left(\frac{1}{0\cdot4} - \frac{1}{0\cdot5}\right)$$

from which

$$\mu = 0 \cdot 236 \frac{\text{(poundal) (sec)}}{\text{ft}^2} \quad \text{or} \quad \frac{\text{lb}}{\text{(ft) (sec)}}$$

This is equivalent to

$$\frac{0 \cdot 236}{0 \cdot 0672} = 3 \cdot 514 \,\text{poise} \quad \text{or} \quad 351 \cdot 4 \,\text{cp}$$

The above approach in calculating viscosity is based on the fact, already observed by Newton, that an infinitely long cylinder rotating around its axis at a constant angular speed will produce in the surrounding fluid

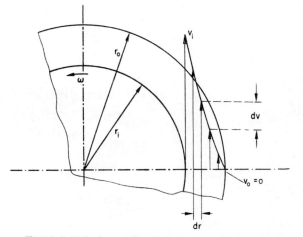

FIG. 1.4. Velocity profile between co-axial cylinders.

a velocity that varies inversely with the radial distance from the axis, as shown in Fig. 1.4.

A simplified approach, based on the assumption that the velocity varies linearly with radius, leads to an error which, however, may not be appreciable in many problems. Basing the calculation on the thus simplified assumption, the procedure is as follows:

$$-\frac{dv}{dy} = \frac{dv}{dr} = \frac{\varDelta v}{\varDelta r} = \frac{v_i - v_o}{r_o - r_i}$$

$$\frac{dv}{dr} = \frac{2 \cdot 512 - 0}{0 \cdot 5 - 0 \cdot 4} = 25 \cdot 12 \,\text{sec}^{-1}$$

Expressing the shear stress in terms of the average radius

$$r = \frac{0 \cdot 4 + 0 \cdot 5}{2} = 0 \cdot 45 \text{ ft}$$

$$\tau = \frac{1 \cdot 187}{r^2} = \frac{1 \cdot 187}{(0 \cdot 45)^2}$$

$$\tau = 5 \cdot 85 \text{ poundal/ft}^2$$

From eq. (1.11)

$$\mu = -\frac{\tau}{dv/dy} = \frac{\tau}{dv/dr} = \frac{5 \cdot 85}{25 \cdot 12}$$

$$\mu = 0 \cdot 233 \frac{\text{(poundal) (sec)}}{\text{ft}^2}$$

equivalent to

$$3 \cdot 465 \text{ poise} \quad \text{or} \quad 346 \cdot 5 \text{ cp}$$

1.9. The Reynolds Number

In a series of papers published in 1883, Osborne Reynolds presented the results of his classical work on the nature of flow in pipes. The work has since become famous for the formulation of a parameter now bearing his name. The parameter takes the form of a dimensionless group of terms and is defined by

$$Re = \frac{Dv\varrho}{\mu} \tag{1.13}$$

where the symbols v, ϱ, and μ refer to fluid in flow through a pipe of diameter D, while Re is the symbol usually given to the so defined parameter, and named the *Reynolds number*.

The formulation contained in eq. (1.13) is the outcome of the most successful correlation of experimental data in a long series of attempts to produce a means of predicting the nature of flow in pipes, and—as will be shown later—in other confinements as well. The significance of the correlation may be appreciated from the fact that—among other things—it enables us to draw a demarcation line between the laminar and non-laminar pattern of flow, a distinction of particular interest in the mathematical analysis of flow phenomena.

Experimental evidence led Reynolds to the specification of a useful number, later called the *critical number*, for the dimensionless group to mark the dividing line between the two patterns of flow. He gave 1400 as his first value, but later corrected this figure to between 1900 and 2000. The latter has since been accepted as the upper limit for laminar flow, and the velocity corresponding to a Reynolds number of 2000 has been specified as the *critical velocity*.

The present-day approach to the Reynolds number, as a criterion of the nature of flow, is more conventional than realistic. Although there is strong evidence that flow in pipes is always laminar for Reynolds numbers up to 2000, there is also a great deal of evidence that fully turbulent flow may not develop until this number exceeds 10,000. It will be recalled at this stage that this kind of flow is characterised by a chaotic movement of fluid particles as distinct from an orderly movement, characteristic of laminar flow.

If 2000 is a reasonable figure for the upper limit in the laminar range, there is definitely less justification in accepting 3000 as the lower limit for the turbulent range, as has been suggested. Nevertheless, these two figures are still widely quoted as the critical Reynolds numbers, the range in between being conveniently accepted as reflecting unstable conditions.

The use of eq. (1.13) is not confined to pipes only. The concept on which the equation is based is a more general one, and finds an extensive application in all possible situations where relative motion between fluid and a solid boundary occurs. The only requirement that may have to be met in any new circumstances refers to the term D contained in eq. (1.13), which must be replaced by another characteristic linear dimension of the solid boundary. In flow past a sphere, for example, the characteristic dimension is its diameter, while in flow through passages of non-circular sections an equivalent diameter is used. It is defined by

$$D_e = 4R_H$$

where R_H is called the *hydraulic radius*, which in turn is defined as the ratio of the flow-area and the wetted perimeter. Thus, for a square cross-section, when b is a side of the square

$$D_e = 4 \left(\frac{b^2}{4b} \right) = b$$

Finally, it will be mentioned that eq. (1.13) may also be presented in forms different from the one given. A commonly used form is

$$Re = \frac{DG}{\mu} \tag{1.14}$$

where $$G = v\varrho \qquad (1.15)$$

The symbol G represents a quantity called the *mass-velocity*. It may be expressed on a second or hourly basis, and care must be exercised that consistent units are adopted in its evaluation.

EXAMPLE 1.5

A liquid of specific gravity 0·9 and viscosity 20 cp flows through the annulus formed by two concentric pipes at a rate of 10 ft/sec. If the inner diameter of the larger pipe is 2 ft and the outer diameter of the smaller pipe is 1·5 ft, what is the Reynolds number corresponding to the flow?

Solution

Let D_o and D_i be the diameters of the larger and smaller pipes, respectively, then the equivalent diameter is

$$D_e = 4R_H = 4\,\frac{(D_o^2 - D_i^2)\,(\pi/4)}{(D_o + D_i)\,(\pi)}$$

$$D_e = D_o - D_i = 2{\cdot}0 - 1{\cdot}5 = 0{\cdot}5 \text{ ft}$$

From eq. (1.15)

$$G = v\varrho = 10 \text{ (ft/sec)}\,(0{\cdot}9)\,(62{\cdot}4)\,(\text{lb/ft}^3)$$
$$G = 561{\cdot}6 \text{ lb/(ft}^2)\,(\text{sec})$$

From eq. (1.14)

$$Re = \frac{D_e G}{\mu} = \frac{(0{\cdot}5)\,(561{\cdot}6)}{(20)\,(0{\cdot}000672)} \; \frac{(\text{ft})\,(\text{lb/ft}^2 \times \text{sec})}{\text{lb/ft} \times \text{sec}}$$

$$Re = 20,890$$

The number is dimensionless, and its high value indicates fully turbulent flow.

1.10. Momentum

The problem of impact attracted considerable attention during the time of Newton. It proved, however, more difficult to solve in those days than it may have seemed today. Galileo was one of the first to tackle impact phenomena competently. He could not, however, have succeeded in solving the problem in principle, as—like some of his contemporaries involved in the idea—he lacked the requisite tools forged later by Newton. The decisive factor in impact is—as we now know—the *momentum*, definable as the product of mass and velocity. This was, however, not conceivable until Wallis, a man of considerable mathematical attainment, clarified the idea in his paper submitted in 1671 to the then newly founded Royal Society of London. The work of Wallis was later supplemented, mainly by the contributions of Wren and Huygens but it was left to Newton to give a concise definition of momentum in relation to the other laws of motion.

The strong interdependence between impact and momentum makes it convenient to approach the theory of momentum from the behaviour of inelastic bodies in collision. The theory is based on the law of conservation of momentum, which may be stated: *for any collision, the algebraic sum of the momenta of the colliding bodies is the same after the impact as before it.*

Consider two bodies, of masses m_1 and m_2, moving linearly in the same direction, the first moving faster. When this body collides with the slower one, the latter will be speeded up and the former slowed down. After the impact, the two bodies will be moving in the same direction with a common velocity. Let this velocity be u, then, following the law of conservation of momentum

$$m_1 v_1 + m_2 v_2 = (m_1 + m_2)u \qquad (1.16)$$

where v_1 and v_2 are the velocities of the faster and slower body respectively, before their collision. Rearranging the terms

$$m_1(v_1 - u) = m_2(u - v_2) \qquad (1.17)$$

Although eq. (1.17) may be regarded as merely a statement of experimental fact, it will be shown to be in agreement with Newton's second law of motion, from which this equation can be derived.

Expressing acceleration as the first derivative of velocity, eq. (1.2) becomes

$$F = m(dv/d\theta) \qquad (1.18)$$

from which

$$F(d\theta) = m(dv) \qquad (1.19)$$

The product of force and time, which appears on the left side of eq. (1.19), is called the *impulse*. The equation then states that the change in momentum of a body equals the impulse given to that body. As a matter of history, the equation was never presented by Newton, but is generally regarded as a mathematical expression of his third law of motion, which states that *every action is met by a reaction.*

Integration of eq. (1.19) between the velocity limits $v = v$, and $v = u$, leads to

$$F\theta = m(v-u) \tag{1.20}$$

which is essentially a momentum equation.

Equation (1.20) finds useful application in problems dealing with impulses and reaction forces produced by jets of fluids. The velocity of a jet

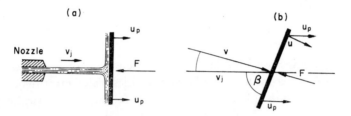

FIG. 1.5. Impact of jets on flat plates.

is normally taken at its mean value, though this approach is not quite consistent with the definition of momentum, which refers to a definite velocity of a single particle. In engineering work, however, this deviation from strictness in procedure is too small to require correction.

Consider a jet issuing from a nozzle at right angles to a flat plate, as shown in Fig. 1.5a. Assume at first that both the nozzle and plate remain stationary, the jet having a velocity v_j. The force produced by the jet, i.e. the thrust on the plate, can be calculated from eq. (1.20). This force is the same as would be required to support the plate from its opposite side in order to keep it stationary.

Taking one second as the basis for the calculation, the mass of fluid striking the plate is

$$\frac{m}{\theta} = a\varrho v_j$$

where a is the cross-sectional area of the jet. For a stationary plate, this mass of fluid is brought to rest at the end of one second, then $u_p = 0$, and the relative velocity is

$$v - u = v_j - u_p = v_j$$

From eq. (1.20)

$$F = \frac{m}{\theta}(v - u) = a\varrho v_j^2$$

Now consider the case when the plate is moving away from the nozzle with its own velocity u_p, while the latter remains stationary as before. The mass of fluid now intercepted by the plate every second is only

$$\frac{m}{\theta} = a\varrho(v_j - u_p)$$

and the relative velocity is now $(v_j - u_p)$, then

$$F = a\varrho(v_j - u_p)^2$$

The same principles apply to the case when the plate is inclined to the jet (Fig. 1.6b), except for the relative velocities which must then be taken at right angles to the plate, as shown in the following numerical example.

EXAMPLE 1.6

A jet of water issuing from a stationary nozzle strikes a flat plate at a velocity of 40 ft/sec. If the jet is inclined to the plate at 30° and is 3 in. in diameter, what is the thrust produced, assuming the plate (*a*) stationary, (*b*) moving with its own velocity of 10 ft/sec in the direction of the jet.

Solution .

Referring to Fig. 1.5b

(a) The mass of water striking the stationary plate is

$$\frac{m}{\theta} = a\varrho v_j = (62 \cdot 4)\left(\frac{\pi}{4}\right)\left(\frac{3}{12}\right)^2 (40)$$

$$\frac{m}{\theta} = 119 \cdot 3 \text{ lb/sec}$$

The relative velocity, taken at right angles to the plate when $u = u_p = 0$, is

$$v - u = v_j \sin \beta = (40)(\sin 30°) = 20 \text{ fps}$$

From eq. (1.20)

$$F = \frac{m}{\theta}(v-u) = (119 \cdot 3)(20)$$

$$F = 2386 \text{ poundals, equivalent to}$$

$$\frac{2386}{32 \cdot 2} = 741 \text{ lb-f}$$

(b) For the plate moving with its own velocity of $u_p = 10$ fps, while the nozzle remains stationary

$$\frac{m}{\theta} = a\varrho(v_j - u_p) = (62 \cdot 4)\left(\frac{3}{12}\right)^2 \left(\frac{\pi}{4}\right)(40-10)$$

$$\frac{m}{\theta} = 91 \cdot 83 \text{ lb/sec}$$

The relative velocity is

$$v-u = (v_j - u_p)\sin\beta = (40-10)(0\cdot5) = 15 \text{ fps}$$

$$F = \frac{m}{\theta}(v-u) = (91\cdot83)(15)$$

$$F = 1378 \text{ poundals, or } 428 \text{ lb-f}$$

1.11. Kinetic Energy and Velocity Head

A body of mass m moving with a velocity u has a kinetic energy of $mu^2/2$. The idea of expressing the quantity of motion in terms of the square of velocity rather than of its first power, as envisaged by momentum, was conceived by Leibniz at the end of the seventeenth century. Introducing the new concept of energy, Leibniz referred to the product mu^2, which he called the *life-force* (*vis-viva*), in contradiction to the *dead-force* (*vis-mortua*), as he preferred to call the Newtonian concept of force. The name *vis-viva* survived until 1807 when Joung proposed a new name—energy, which was later changed to dynamic energy, and finally to *kinetic energy*, as suggested by Kelvin in 1856, to distingish this form of energy from another one which does not involve motion, and which is now known as *potential energy*.

In the meantime (1835), a French physicist (Coriolis) pointed out that the concept of kinetic energy could more justifiably be applied to one-half the product of mass and square of velocity, as this quantity could be directly derived from Newton's second law of motion. This suggestion has since received general recognition, and the expression $\frac{1}{2}mu^2$ has become the measure of the kinetic energy resident in moving bodies. (Only a passing reference will be made here to the fact that this quantity loses its validity for very high velocities, in which case Einstein's theory of relativity plays a decisive role.)

For real fluids in motion, kinetic energy is usually expressed in terms of the average velocity, defined by $v = Q/A$, where Q is the volumetric rate of flow and A the cross-section of a streamtube, or more commonly the cross-section of a pipe. The expression takes the form

$$\text{K.E.} = \frac{mv^2}{2\alpha} \tag{1.21}$$

in which α is a correcting factor to allow for the velocity distribution across a section. For a real fluid, the distribution depends on the pattern of flow, as shown in Fig. 1.6.

For laminar flow, the velocity distribution is parabolic, and independent of the Reynolds number up to its critical value of 2000. Within this range, the average velocity is exactly half the maximum velocity at the axis of the pipe, and also $\alpha = 0.5$. There is an increase in both with increase in the average velocity, the limiting value being $\alpha = 1$ for $Re = \infty$. This limiting condition is never met in practice but can be fairly closely approached by fast flowing gases. Flow of this kind is called *plug flow*. It is also characteristic of the *ideal flow*, i.e. the flow of an ideal fluid. The variation of the correcting factor α with the Reynolds number is shown in Fig. 1.7.

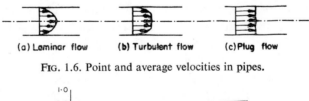

(a) Laminar flow   (b) Turbulent flow   (c) Plug flow

FIG. 1.6. Point and average velocities in pipes.

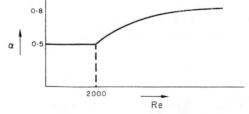

FIG. 1.7. The change of velocity profile with the type of flow.

In the subject of fluid dynamics, kinetic energy is more conveniently expressed in terms of the so-called *velocity head*. This is based on unit mass ($m = 1$) of fluid in motion, and defined by the equation

$$H_v = \frac{v^2}{2\alpha g} \tag{1.21a}$$

Dimensionally, this expression has a unit of length, hence the name *head* given to it.

In streamline motion ($Re < 2000$), $\alpha = 0\cdot5$, and the expression reduces to

$$H_v = \frac{v^2}{g}$$

The correcting factor α increases rapidly above the critical value of the Reynolds number of 2000, and reaches a value in excess of $0\cdot8$ for flows normally encountered in industrial practice. Being fairly close to unity, this factor is usually ignored in calculations, and the *velocity head* is then commonly written

$$H_v = \frac{v^2}{2g} \tag{1.22}$$

As a matter of interest, the value of $0\cdot5$ attached to α, within the whole laminar range of flow, can be proved as follows.

Consider the flow of a real fluid in a pipe of radius r, and assume its cross-section to be divided into a number of annular sections of differential area, dA each, as shown in Fig. 1.8. Since the sections are formed by

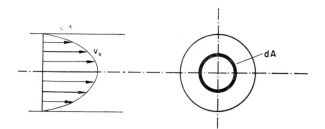

FIG. 1.8. The correcting factor (α) as a function of (Re).

narrow circular strips, the velocity of fluid passing through each section may be accepted to be uniform. Let this velocity be v_x at a radial distance y, then the fluid of differential mass dm passing through a section will have a kinetic energy of

$$d(\text{K.E.}) = (dm)(v_x^2/2)$$

Since $dm = \varrho v_x(dA)$, then this elemental quantity of kinetic energy is $\varrho v_x(dA)\, v_x^2/2$, or

$$d(\text{K.E.}) = \tfrac{1}{2}\varrho(dA)\, v_x^3$$

For a circular section of radius y, $A = \pi y^2$, and $dA = 2\pi y(dy)$, then

$$d(\text{K.E.}) = \pi\varrho v_x^3 y(dy)$$

For a pipe of radius r

$$\text{K.E.} = \pi\varrho \int_0^r v_x^3 y(dy) \tag{1}$$

where v_x is also some function of y.

Expressing the kinetic energy in terms of the average velocity of flow

$$\text{K.E.} = \frac{mv^2}{2\alpha} \tag{1.21}$$

The mass of the fluid passing every second through the pipe is

$$m = vr^2\pi\varrho$$

Substitution gives

$$\text{K.E.} = \frac{\pi\varrho v^3 r^2}{2\alpha} \tag{2}$$

Equating (1) and (2)

$$\frac{\pi\varrho v^3 r^2}{2\alpha} = \pi\varrho \int_0^r v_x^3 y(dy)$$

from which

$$\alpha = \frac{v^3 r^2}{2\int v_x^3 y(dy)}$$

For streamline motion, from eq. (3.54) (see Chapter 3)

$$v_x = 2v\left[1 - \left(\frac{y}{r}\right)^2\right]$$

It follows that

$$\alpha = \frac{v^3 r^2}{2 \int 8v^3 \left[1 - \left(\frac{y}{r}\right)^2\right]^3 y \, (dy)}$$

$$\alpha = \frac{r^2}{16 \int \left[1 - \left(\frac{y}{r}\right)^2\right]^3 y \, (dy)} \tag{3}$$

The integral of this equation

$$\int \left[1 - \left(\frac{y}{r}\right)^2\right]^3 y \, (dy) = \frac{1}{r^6} \int (r^2 - y^2)^3 y \, (dy)$$

$$= \frac{1}{r^6} \int (r^6 - 3r^4 y^2 + 3r^2 y^4 - y^6) \, y (dy)$$

$$= \frac{1}{r^6} \left[\frac{y^2 r^6}{2} - \frac{3r^4 y^4}{4} + \frac{3r^2 y^6}{6} - \frac{y^8}{8}\right]_{y=0}^{y=r}$$

$$= \frac{r^2}{8}$$

Substituting in eq. (3)

$$\alpha = \frac{r^2}{16(r^2/8)}$$

$$\alpha = 0 \cdot 5 \tag{4}$$

1.12. The Equation of Continuity

The following equation applies to fluid in steady motion

$$\varrho v A = \text{constant} \tag{1.23}$$

Based on the law of conservation of mass, the equation of continuity may be applied to any number of sections of a streamtube, or pipe, with the effect that the resulting equation takes the form of a simple material balance.

Consider the steady flow through a streamtube, as shown in Fig. 1.9. When applied to sections *1* and *2* of the tube, the continuity equation becomes

$$\varrho_1 v_1 A_1 = \varrho_2 v_2 A_2 \tag{1.24}$$

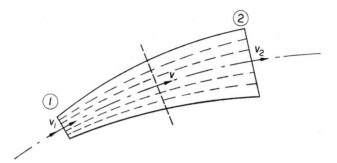

FIG. 1.9. The application of the law of continuity to a streamtube.

where the symbols A_1 and A_2 represent the cross-sectional areas at the two sections considered, measured in planes normal to the axis of the tube.

The alternative forms of eq. (1.24) are

$$\varrho_1 Q_1 = \varrho_2 Q_2 \qquad (1.25)$$
$$G_1 A_1 = G_2 A_2 \qquad (1.26)$$

where the symbol G represents a quantity known as the *mass-velocity*, as defined by eq. (1.15).

1.13. Bernoulli's Theorem

In an isolated system, the form and availability of energy may change but the total energy remains constant. This statement is a general way of expressing *the law of conservation of energy*. When applied to the subject of fluid dynamics, the law is referred to as *Bernoulli's theorem*.

The theorem refers in principle to a single fluid particle in steady flow along a streamline. Nevertheless, it can be equally well applied to the flow of an ideal fluid across a section of a streamtube, or pipe.

Consider a section of a streamtube through which an ideal fluid passes at a uniform velocity v, such as the section in the plane x–x, shown in Fig. 1.10. Let m be the mass of the fluid crossing the section at any time and V' be the volume in which this mass is contained, under the conditions existing at the section. For the sake of simplicity, assume that this mass is concentrated at the point of intersection of the section with the *axis* of the tube. If Z is the vertical distance of this point from a reference line, such as the o–o line, then the total energy carried by the mass considered

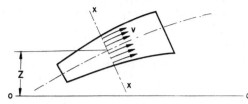

Fig. 1.10. Flow of an ideal fluid through a streamtube.

is the sum of the energies due to the static pressure, velocity, and location above the reference line. Since the total energy is the same at all sections of the tube, then for any section the condition is met by the equation

$$pV' + \frac{mv^2}{2} + mZg = \text{constant} \tag{1.27}$$

in which each term is expressed in energy units.

In Fluid Mechanics, it is convenient to deal with a unit mass of the flowing fluid. On this basis, eq. (1.28) is obtained by dividing the term of eq. (1.27) by m

$$\frac{pV'}{m} + \frac{v^2}{2} + Zg = \text{constant} \tag{1.28}$$

Multiplying the last equation by

$$\varrho = \frac{m}{V'}$$

$$p + \frac{\varrho v^2}{2} + \varrho Zg = \text{constant} \tag{1.29}$$

Each term of this equation is expressed in absolute units of pressure, and the form presented by eq. (1.29) is usually referred to as the Bernoulli equation.

In engineering practice, it is often preferred to express pressure in the gravitational system of units. To meet this requirement, eq. (1.29) is written

$$pg_c + \frac{\varrho v^2}{2} + \varrho Zg = \text{constant} \tag{1.30}$$

where g_c is the gravitational conversion factor, as defined by eq. (1.3). Dividing the terms of the last equation by ϱg

$$\frac{pg_c}{\varrho g} + \frac{v^2}{2g} + Z = \text{constant} \tag{1.31}$$

The terms of this equation are expressed in units of length. For this reason, the three terms are referred to as *pressure head*, *velocity head* and *potential head*, respectively.

The equation is more commonly presented in the form

$$\frac{p}{\gamma} + \frac{v^2}{2g} + Z = \text{constant} \tag{1.32}$$

in which $\gamma = \varrho(g/g_c)$, as defined by eq. (1.6).

One more form of the Bernoulli equation is obtained by multiplying the terms of eq. (1.31) by g/g_c. The resulting equation is

$$\frac{p}{\varrho} + \frac{v^2}{2g_c} + Z(g/g_c) = \text{constant} \tag{1.33}$$

In eq. (1.33) pressure is expressed in engineering units while absolute units are used for density. This calls for special attention in dealing with problems which make use of this equation.

EXAMPLE 1.7

A liquid, of density 60 lb/ft³, is conveyed up a conical tube at the rate of 3·14 ft³/sec. The tube is 6 in. diameter at the lower end and 12 in. at the upper end, and the two ends are separated by a vertical distance of 10 ft. If the pressure at the lower end is 50 psia, what is the pressure at the upper end? Assume the liquid to behave like an ideal fluid.

Solution

Applying the law of continuity to the end sections

$$Q = v_1 A_1 = v_2 A_2 = 3\cdot14 \text{ cfs}$$
$$A_1 = (6/12)^2 (\pi/4) = \pi/16$$
$$A_2 = (12/12)^2 (\pi/4) = \pi/4$$
$$v_1 = \frac{(3\cdot14)(16)}{\pi} = 16 \text{ fps}$$
$$v_2 = \frac{(3\cdot14)(4)}{\pi} = 4 \text{ fps}$$

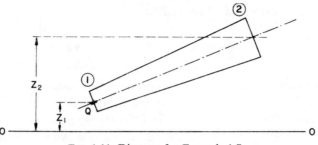

FIG. 1.11. Diagram for Example 1.7.

Bernoulli's theorem, when applied between the two sections, using eq. (1.32), leads to

$$\frac{p_1}{\gamma}+\frac{v_1^2}{2g}+Z_1 = \frac{p_2}{\gamma}+\frac{v_2^2}{2g}+Z_2$$

from which

$$\frac{p_2-p_1}{\gamma} = \frac{v_1^2-v_2^2}{2g}+(Z_1-Z_2)$$

Substituting in this equation the data, and noting that

$$p_2-p_1 = (p_2-50)\,144 \text{ psf}$$

$$\gamma = 60\,\frac{\text{lb-f} = \text{lb-wt}}{\text{ft}^3}$$

$$2g = (2)(32 \cdot 2) = 64 \cdot 4 \text{ ft/sec}^2$$

$$Z_1-Z_2 = -(Z_2-Z_1) = -10 \text{ ft}$$

$$\frac{(p_2-50)144}{60} = \frac{16^2-4^2}{64 \cdot 4}-10 = -\frac{101}{16 \cdot 1}$$

$$p_2-50 = -\frac{(101)(60)}{(16 \cdot 1)(144)} = -2 \cdot 614, \text{ say } -2 \cdot 6$$

$$p_2 = 47 \cdot 4 \text{ psia}$$

1.14. The Flow Equation

The Bernoulli equation can be modified to cover all possible situations in which the law of conservation of energy is applicable. A modified form which takes into account heat effects and external work, and also energy losses in the flow of real fluids is called the *flow equation*. The following form is an extension of eq. (1.22):

$$\frac{p}{\gamma}+\frac{v^2}{2\alpha g}+Z+q+W+H_f = \text{constant} \qquad (1.34)$$

The conversion factor α which appears in this equation allows for the departure from ideal flow, as explained in an earlier paragraph. Unless the flow is laminar, in which case $\alpha = 0\cdot5$, the conversion factor is ignored in most engineering problems. The remaining new symbols which appear in this equation, q, W, and H_f, represent heat, work, and energy losses, respectively. The first two may be assigned negative or positive values, depending on whether they represent the output from or input to the flow system, unlike the H_f term which must always appear on the output side of the flow equation with a positive sign. The methods of evaluation of this term will be dealt with in later chapters.

EXAMPLE 1.8

Sulphuric acid (sp.gr. $= 1\cdot8$) is to be pumped from an open tank to a process column at the rate of 18 lb/sec. The column operates at 19·65 psia, and the acid is sprayed into it from a nozzle situated 60 ft above the acid surface level in the tank, with a velocity of 8 fps. If the energy losses are estimated to be equivalent to 9 ft water head, what power will be required to run the pump with an overall efficiency of 60 per cent? The barometer reading is 750 mm.

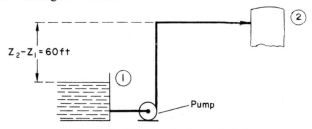

FIG. 1.12. Diagram for Example 1.8.

Solution

Applying eq. (1.34) between the two sections shown in Fig. 1.12

$$\frac{p_1}{\gamma}+\frac{v_1^2}{2\alpha g}+Z_1+W+q = \frac{p_2}{\gamma}+\frac{v_2^2}{2\alpha g}+Z_2+H_f \qquad (1.35)$$

In the absence of heat effects, $q = 0$, and for a large surface in the tank, $v_1 = 0$, and also taking $\alpha = 1$, eq. (1.35) reduces to

$$W = (Z_2 - Z_1) + \frac{p_2 - p_1}{\gamma} + \frac{v_2^2}{2g} + H_f$$

where W represents the net work to be done by the pump on each pound of the acid conveyed. The right-hand-side terms of this equation have the following values in terms of the *acid head*:

$$Z_2 - Z_1 = 60\cdot0 \text{ ft}$$
$$v_2^2/2g = 8^2/(2)(32\cdot2) = 1\cdot0 \text{ ft}$$
$$H_f = \frac{9\cdot0 \text{ (ft water)}}{1\cdot8} = 5\cdot0 \text{ ft}$$

Taking $p_1 = \dfrac{750}{760}(14\cdot7) = 14\cdot65$ psia

$$\frac{p_2 - p_1}{\gamma} = \frac{(19\cdot65 - 14\cdot65)(144)}{(1\cdot8)(62\cdot4)} = 6\cdot4 \text{ ft}$$
$$W = 60 + 6\cdot4 + 1\cdot0 + 5\cdot0$$
$$W = 72\cdot4 \text{ ft}$$

Making use of the concept of weight, 1 lb-f = 1 lb-wt, this work can be expressed as

$$W = 72\cdot4 \frac{\text{(lb-f)(ft)}}{\text{(lb-wt)}}$$

The work to be done in pumping 18 lb of the acid per second is

$$72\cdot4 \frac{\text{(lb-f)(ft)}}{\text{(lb-wt)}} \times 18 \frac{\text{(lb-wt)}}{\text{sec}} = (72\cdot4)(18) \text{ (lb-f)(ft)/sec}$$

The power required to run the pump with an overall efficiency of 60 per cent is then

$$\frac{(72\cdot4)(18) \text{ (lb-f)(ft)/sec}}{(0\cdot6)(550) \text{ (lb-f)(ft)/sec}} = 3\cdot95 \text{ hp}$$

1.15. Hydraulic Gradient

For an ideal flow, eq. (1.32) may be presented in the form

$$Z + \frac{p}{\gamma} + \frac{v^2}{2g} = H_o \tag{1.36}$$

in which H_o is the total energy carried by a unit mass of ideal fluid in motion. The contribution of the three left-hand-side terms of this equation towards the total energy is shown in the diagram below.

In ideal flow, there is no energy loss, and the total energy H_o is the same at any section of a streamtube, as shown in Fig. 1.13. Consequently the plot of the velocity, pressure and potential head results in a horizontal line, called the *total energy line*. The plot of the sum of the pressure and potential energy only gives a line called the *hydraulic grade line*, or simply the *hydraulic line*. This line shows the level to which the liquid would rise in a vertical stand pipe at the sections under consideration.

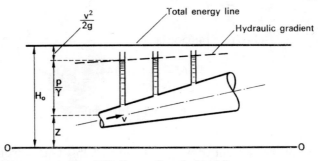

FIG. 1.13. Energy lines in ideal flow.

In real flow, the total energy line and the hydraulic gradient are affected by energy losses. If the heat and work terms of eq. (1.34) are ignored, the equation may be presented in the form

$$\frac{p}{\gamma} + \frac{v^2}{2\alpha g} + Z = H_o - H_f = H \tag{1.37}$$

in which the total energy H carried by a unit mass of fluid is less than that available in ideal flow by an amount H_f lost in friction. The effect of friction on the total energy and hydraulic lines is illustrated in Fig. 1.14, which shows the total energy and hydraulic lines in the flow between two tanks when the total head available is due to the difference between the surface levels in the tanks. The losses are mainly due to friction in the lines, but small losses are also be to accounted for due to changes in sections, such as between pipes of different diameters, and between pipes and tanks. These losses have, however, been ignored at this stage.

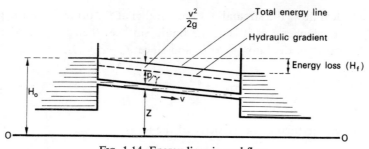

FIG. 1.14. Energy lines in real flow.

1.16. The Coefficient of Discharge

The energy loss in the flow of a real fluid can often be conveniently expressed in terms of some coefficient. This particularly applies to losses in streamtubes or passages which are difficult to define geometrically, such as formed by orifices and pipe fittings. Of special interest in Fluid Dynamics is the *coefficient of discharge*. The definition of this coefficient and its relation to some other coefficients is dealt with in connection with flow through small orifices.

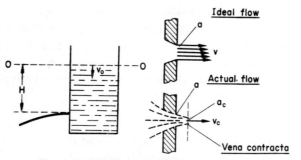

FIG. 1.15. Discharge through a small orifice.

Consider, in Fig. 1.15, a jet of liquid issuing from a small orifice, located in the side of a large tank filled with the liquid to a height H, measured above the centre of the orifice. Assuming ideal flow, the velocity of the jet can be related to the liquid head in the tank by the application of Bernoulli's theorem between the liquid surface in the tank (level o–o), and the centre of the orifice. Let this velocity be v, then from eq. (1.32)

$$\frac{p_o}{\gamma} + \frac{v_o^2}{2g} + Z_o = \frac{p}{\gamma} + \frac{v^2}{2g} + Z \qquad (1.38)$$

In an ideal flow, the streamlines are parallel at the orifice, and the pressure p is atmospheric as it is at the liquid surface: then $p = p_o$. Also for a large section in the tank, $v_o = 0$, and since $Z_o - Z = H$, then eq. (1.38) reduces to

$$H = \frac{v^2}{2g} \qquad (1.39)$$

from which the theoretical velocity of the jet is

$$v = \sqrt{(2gH)} \qquad (1.40)$$

This equation is a mathematical formulation of the statement made by Torricelli.

Let a be the cross-sectional area of the orifice. For ideal flow, this is also the area of the jet itself, hence the theoretical discharge under the head H is

$$Q_t = a\sqrt{(2gH)} \qquad (1.41)$$

The energy loss experienced in real flow across the orifice makes the actual discharge considerably less than the theoretical one. In the flow of a real fluid, the streamlines converge at the orifice with the effect that fluid particles have a vertical velocity component in addition to the horizontal one. This component disappears only a little distance outside the orifice, and the section of the jet at which the streamlines first become parallel is called the *vena contracta*. Let a_c and v_c be the cross-sectional area and average velocity of the jet at this section respectively, then the actual discharge is

$$Q = a_c v_c$$

Now let $a_c = C_c a$, and $v_c = C_v v$, where C_c and C_v are the coefficients of *contraction* and *velocity*, respectively. It follows that

$$Q = (C_c)(C_v)\,av$$

Again let

$$C_D = (C_c)(C_v)$$

then

$$Q = C_D av$$

Substituting for v from eq. (1.40)

$$Q = C_D a\sqrt{(2gH)} \qquad (1.42)$$

where C_D is the *coefficient of discharge*.

If h_f be the head lost in the flow, then the net head available for this flow is $H-h_f$, and the actual rate of flow is also given by

$$Q = a\sqrt{\{2g(H-h_f)\}} \tag{1.43}$$

From this equation, and eq. (1.42),

$$C_D\sqrt{H} = \sqrt{(H-h_f)}$$

from which

$$h_f = H(1-C_D^2) \tag{1.44}$$

The coefficient of discharge is normally determined experimentally, but it can also be obtained from values of the coefficients of velocity and contraction. The latter can be measured while the coefficient of velocity can be obtained indirectly from experimental data, as follows.

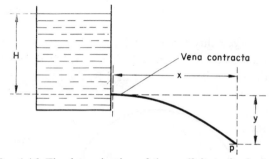

FIG. 1.16. The determination of the coefficient of velocity.

Figure 1.16 shows a jet intercepted at a point p. Let x be the horizontal distance of the point from the vena contracta and y the vertical distance from the axis of the orifice. From these two measurements, the coefficient of velocity can be calculated from eq. (1.48), the derivation of which is given below.

Let θ be the time required for a fluid particle to travel from the vena contracta to the point p, then resolving the velocity into its horizontal and vertical components

$$x = v_c\theta \tag{1.45}$$

$$y = g\theta^2/2 \tag{1.46}$$

where v_c is the average velocity of the jet at the vena contracta and g is the acceleration due to gravity. Solving the two equations simultaneously

$$y = \frac{g}{2}\left(\frac{x}{v_c}\right)^2$$

from which

$$v_c = \sqrt{\left(\frac{gx^2}{2y}\right)} \qquad (1.47)$$

But

$$v_c = C_v v = C_v\sqrt{(2gH)}$$

hence

$$C_v\sqrt{(2gH)} = \sqrt{\left(\frac{gx^2}{2y}\right)}$$

from which

$$C_v = \frac{x}{\sqrt{(4yH)}} \qquad (1.48)$$

EXAMPLE 1.9

Water is discharged through an orifice of diameter 1 in. at the bottom of a closed tank, as shown in Fig. 1.17. In a test, 351 lb is collected in one minute when the depth is maintained at 7 ft, the space above the surface

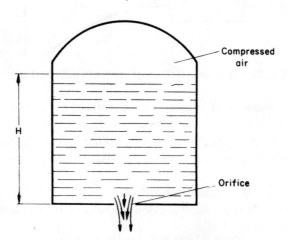

FIG. 1.17. Diagram for Example 1.9.

of the water being kept at a pressure of 2·17 psig by compressed air. If the measured diameter of the jet at the vena contracta is 0·8 in., what are the coefficients of *contraction, velocity* and *discharge* under the conditions of the test? What is the loss of head, expressed in feet of water?

Solution

$$\text{Coefficient of contraction} = \frac{\text{Jet area}}{\text{Orifice area}}$$

$$C_c = \frac{a_c}{a} = \left(\frac{0\cdot8}{1\cdot0}\right)^2 = 0\cdot64$$

The rate of discharge in cfs is

$$Q = \frac{351}{(60)(62\cdot4)}$$

The total head available

$$H + p/\gamma = 7 + \frac{(2\cdot17)(144)}{62\cdot4} = 12 \text{ ft}$$

From eq. (1.42)

$$Q = C_D a \sqrt{\{2g(H+P/\gamma)\}} = C_D \left(\frac{1}{12}\right)^2 \left(\frac{\pi}{4}\right) \sqrt{\{(64\cdot4)(12)\}}$$

Equating

$$\frac{351}{(60)(62\cdot4)} = C_D \left(\frac{\pi}{568}\right) \sqrt{772\cdot8}$$

from which

$$C_D = 0\cdot61$$

From eq. (1.44)

$$h_f = (H+P/\gamma)(1-C_D^2) = (7+5)[1-(0\cdot61)^2]$$
$$h_f = 7\cdot536 \text{ ft}$$

As a check, using eq. (1.43), the rate of discharge is

$$60Q\varrho = 60\left(\frac{1}{12}\right)^2\left(\frac{\pi}{4}\right)\sqrt{\{64\cdot4(12-7\cdot536)\}}(62\cdot4) = 351 \text{ lb/min}$$

The coefficient of velocity

$$C_v = \frac{C_D}{C_c} = \frac{0\cdot61}{0\cdot64} = 0\cdot95$$

EXAMPLE 1.10

A jet issues from an orifice located in the vertical side of a tank standing on horizontal ground. If the tank is filled with a liquid of density ϱ to a height h ft, and the centre of the orifice is H ft below the surface of the liquid, what is the value of H if the jet strikes the ground at a maximum distance from the tank? If the orifice has an area of a ft², what is the horizontal reaction of the jet on the tank?

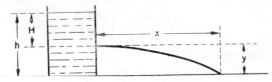

FIG. 1.18. Diagram for Example 1.10.

Solution

From eq. (1.48)

$$x = C_v \sqrt{(4yH)}$$

but $y = h - H$, then

$$x = C_v \sqrt{\{4(h-H)H\}}$$

With reference to Fig. 1.18, x is a maximum when $4(h-H)H$ is a maximum, or when

$$\frac{d[(h-H)H]}{dH} = h - 2H = 0$$

from which

$$H = h/2$$

From eq. (1.42), the volumetric rate of discharge through the orifice is

$$Q = C_D a \sqrt{(2gH)}$$

and the corresponding mass rate of discharge is

$$\varrho Q = \frac{m}{\theta} = \varrho C_D a \sqrt{(2gH)}$$

From eq. (1.22), taking $u = 0$, and $v = v_c$

$$F = \frac{mv_c}{\theta}$$

But $v_c = C_v \sqrt{(2gH)}$, so that the reaction on the stationary tank is

$$F = C_D C_v a (2gH) \varrho$$

1.17. The Concept of the Boundary Layer and its Significance

Fluid in motion relative to a solid boundary develops a velocity gradient in a plane perpendicular to the boundary. The gradient extents from the boundary to a point in the main stream of the fluid body where its magnitude is reduced to a negligible extent. The layer of the fluid contained within this gradient is called the *boundary layer*. As the velocity gradient does not die out abruptly in the main stream, the boundary of this gradient is more definitely defined by the thickness of the layer extending to a point where the velocity is only 1 or 0·1 per cent of the velocity in the main stream, assumed uniform. This layer is always divided into two kinds,

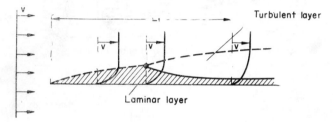

FIG. 1.19. Formation of the boundary layer at flat plates.

one—*laminar*—where viscous stresses are predominant, and another—*turbulent*—where viscosity effects are insignificant.

The boundary layer does not form immediately as fluid approaches a boundary. It builds up gradually along the boundary from zero thickness at the entry point to a maximum, some distance from this point. In the following discussion, this distance will be assigned the symbol L_t.

A simple example of the formation of the boundary layer in flow along a flat plate is shown diagrammatically in Fig. 1.19. The fluid approaching the plate is split into two streams, and the formation of the boundary layer begins at the front edge of the plate, on both sides of it. Considering the upper stream only, the boundary layer is seen to grow in thickness along the plate until it has developed to its full size. Near the origin of its formation, the flow in the layer is wholly laminar, but a little distance further downstream a transition to turbulent flow is observed, with a thin layer closest to the solid surface preserving its laminar state.

Figure 1.20 shows the formation of the boundary layer in a pipe. The *transition length* (L_t), i.e. the length of the pipe required for the boundary layer to develop, depends on the average velocity of flow in the pipe, as

it does in flow along the flat plate discussed earlier. This length may therefore be expressed in terms of the Reynolds number, based on the average
velocity, the relevant equation being

$$\frac{L_t}{D} = kRe \qquad (1.49)$$

where D is the pipe diameter and k a constant. This constant has been
found experimentally to have a value of 0·05 for laminar flow and 0·0288
for turbulent flow.

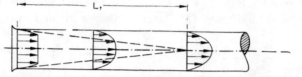

Fig. 1.20.   Formation of the boundary layer in pipes.

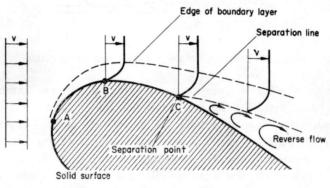

Fig. 1.21. Separation of the boundary layer.

In certain circumstances, the boundary layer may separate from a solid
surface. The phenomenon may be explained with reference to the flow
around a curved surface as follows.

Consider incompressible flow around the curved surface in Fig. 1.21.
The fluid approaching the surface is subjected to acceleration in the region
from the *stagnation point* at A to the highest point of the surface at B,
due to the reduction in the flow area. Following Bernoulli's theorem, the
pressure will decrease in this region in the direction of flow. The reverse
is true in the downstream region from B where the flow area and pressure
increase. A pressure variation of this kind is called an *adverse pressure*

gradient. It has a much lesser effect on the main stream than on the slowly moving boundary layer. The sensivity of this layer to the retarding forces produced by the increased pressure is such that it may eventually be brought to rest and further to reverse the direction of its motion. The point on the surface, such as the point *C* in Fig. 1.21, at which this situation first arises, marks the separation line between the boundary layer and the relatively stationary fluid below it. An outcome of such a separation is discussed below.

Consider in Fig. 1.22 three differently shaped bodies, but all having the same cross-sectional area, immersed in the same fluid. Let the fluid flow past the bodies at a velocity which will produce no separation at the rear

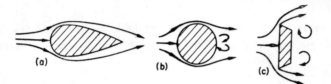

FIG. 1.22. Flow past (a) streamlined body, (b) sphere, and (c) circular disc.

of the well-streamlined body (a). The flow around this body is assumed to be perfectly smooth, with the result that any friction is entirely due to tangential stresses. This kind of friction is traditionally called *skin-friction*. The flow past the remaining two shapes, namely the sphere (b) and circular disk (c), shows separation of the boundary layer with the effect that a *wake* is produced at the rear of the solid boundary. The wake is characterised by large-scale eddies and *vortices*, which require considerable energy to support them. This energy is drawn from the main stream, thus reducing considerably its energy level. This additional frictional effect is due to the so-called *form-friction*, or *form-drag*, the name indicating that its source is derived from the geometry of the obstructing body. As the form drag has a much more adverse effect on flow than the *skin-drag*, a great deal of attention is paid in design practice to eliminate it or, at least, to reduce its effect by streamlining all obstructions to flow. Occasionally, however, the *form-drag* is deliberately introduced to flow systems in order to create some desirable differential effects. An orifice-meter may be cited here as an example.

D'Arcy (1803–58), and Froude (1810–79) had some appreciation of the role played by the boundary layer in the resistance to flow past solid boundaries, but it is Prandtl (1875–1953) who has gained general recognition as the pioneer in the field of modern fluid mechanics, largely based on the concept of the boundary layer. His new approach to the analysis of frictional phenomena, and his original experimental techniques have provided a pattern for research workers engaged in this field to follow. Among many of his followers, Blasius (1883–) should be mentioned for his solution of a differential equation which gives the velocity distribution in the laminar boundary layer along an infinitely large plate in an otherwise uniform stream. But, by far the most succesful of Prandtl's followers was von Karman (1881–1963). His best-known contribution to the subject of the boundary layer is the mathematical analysis of the skin and form drag in pipes and along plane surfaces. This has led him to the logarithmic expressions which bear his name jointly with that of Prandtl.

A number of equations are now available for use with the boundary layer. These are discussed fully in texts[2, 3] and papers devoted to this subject.

1.18. Dimensional Analysis

Any fundamental equation can often be more conveniently expressed in terms of some dimensionless groups of terms rather than in terms of individual variables. The method of organising variables into such groups is called *dimensional analysis*.

Fourier (1768–1830) is supposed to have laid the foundations for what is now known as dimensional analysis as early as 1822. Many others thereafter paid casual attention to this subject, but it was not till 1899 that Rayleigh (1842–1919) generalised the principle involved in the analysis and made it popular not only in fluid mechanics but also in other allied disciplines. Further progress was made with the publication of papers on dimensional analysis by Buckingham (1867–1940), in 1914–15, who presented a new method of formulating non-dimensional parameters by means of what he called the *π-theorem*.

Rayleigh's method, which will be outlined in the following example, is based on the principle that any physical quantity may be expressed as a product of powers of a very few fundamental dimensions. The application of the principle may be simply explained with reference to flow through orifices.

Let v be the velocity of flow of a fluid through an orifice. Let us also assume that this velocity is some function of the fluid head H, and of the gravitational acceleration g. This can be written in the following functional form:

$$v = \phi(H, g)$$

Following Rayleigh's method, the following exponential equation can be set up

$$v = k(H)^x (g)^y \qquad (1.50)$$

in which k is a proportionality constant.

Substituting the dimensions for each variable of this equation

$$(L\theta^{-1}) = k(L)^x (L\theta^{-2})^y$$

If the equation is to be dimensionally consistent, the sum of the exponents on each dimension must be the same on both sides of the equation. The two dimensions can therefore be separated into equations for L and θ, giving for L

$$1 = x + y$$

and for θ

$$-1 = -2y$$

Solving the last two equations simultaneously

$$y = \tfrac{1}{2}, \quad \text{and} \quad x = \tfrac{1}{2}$$

Substituting these values for x and y in eq. (1.50)

$$v = k(H)^{1/2} (g)^{1/2}$$

or

$$v = k \sqrt{(gH)} \qquad (1.51)$$

from which

$$H = \frac{v^2}{k^2 g} \qquad (1.52)$$

The last equation will be recognised as the basic equation for orifices, given earlier in the form

$$H = \frac{v^2}{2g} \qquad (1.39)$$

It follows that in ideal flow the constant $k = \sqrt{2}$.

The evaluation of constants obtained in dimensional analysis is possible by further analytical investigation. Experimental data is, however, necessary for their determination in real flow.

In more complicated problems there may be more unknowns involved than equations available. In such cases the equations can be solved in terms of some arbitrarily selected variables, as shown in the following example. For more comprehensive study of the subject the work of Focken[4] should be consulted.

EXAMPLE 1.11

The volumetric rate of discharge (Q) from a centrifugal pump of diameter (D) may be related to the speed of rotation (N) by the equation

$$Q = kND^3 \left[\frac{p}{\varrho N^2 D^2} \right]^t$$

in which (ϱ) is the density of the liquid discharged, and (p) is the pressure against which the pump operates. How can this equation be obtained by the application of dimensional analysis?

Solution

On the basis of the assumptions made

$$Q = k[\varrho]^x [N]^y [D]^z [p]^t$$

where k is a proportionality constant.

Substituting the dimensions for each variable of the equation, and noting that $[1/\theta]$ is the dimension for $[N]$

$$[L^3\theta^{-1}] = [ML^{-3}]^x [\theta^{-1}]^y [L]^z [ML^{-1}\theta^{-2}]^t$$

Equating the indices for

$$
\begin{array}{cl}
M & 0 = x + t \\
\theta & -1 = -y - 2t \\
L & 3 = 3x + z - t
\end{array}
$$

Solving the resulting equations in terms of t

$$
\begin{array}{l}
x = -t \\
y = 1 - 2t \\
z = 3 - 2t
\end{array}
$$

Substituting from these equations for x, y and z in the original equation

$$Q = k[\varrho]^{-t} [N]^{1-2t} [D]^{3-2t} [p]^{t}$$

Rearranging the terms of this equation

$$Q = kND^3 \left[\frac{p}{\varrho N^2 D^2} \right]^t$$

Problems

1.1. What is the numerical value of the universal gas constant expressed in (atm) (litre)/(gm-mole) (°C), taking 22·4 litres for 1 gm-mole at STP?

Hydrogen (mol.wt. = 2) is under a pressure of 8 atm. What is its density, expressed in gm/cc, at 27°C? (*Ans.*: 0·082, 0·00065)

1.2. Carbon dioxide (mol.wt. = 44) is compressed into a cylinder at 80°F. If its density is 0·304 lb/ft³, what is the pressure in the cylinder, expressed in psia? (*Ans.*: 40)

1.3. A liquid has a kinematic viscosity of 0·052 cm²/sec. If the density of the liquid is 60 lb/ft³, what is its dynamic viscosity, expressed in centipoises?

The liquid fills a space between two parallel plates, 0·1 cm apart. What force, expressed in dynes and lb-f, is required in moving one plate past the other stationary plate, with a velocity of 0·2 cm/sec, if the moving plate has an area of 10,000 cm²? (*Ans.*: 5, 1000, 0·00225)

1.4. A cylinder, of 0·4 ft radius, rotates coaxially inside a stationary cylinder of 0·42 ft radius with a liquid of 2·4 poise viscosity filling the annular space. Assuming linear velocity distribution, what torque, expressed in ft×lb-f, will be required in rotating the cylinder at 60 rpm? The cylinders have a length of 2 ft. (*Ans.*: 1·33)

1.5. A shaft of diameter 2·99 in. rotates in a bearing of 3·00 in. diameter and of 6 in. length. If the annular space is filled with a lubricating oil of 60 centipoises viscosity what force (in poundals), and horse-power is wasted in overcoming resistance in the bearing when the shaft rotates at 600 rpm? Assume linear velocity distribution. (*Ans.*: 296, 0·13)

1.6. What is the hydraulic radius of a rectangular duct of sides 6 and 9 in.? If a gas, of 0·017 cp viscosity and 0·147 lb/ft³ density, flows through the duct at a Reynolds number of 2000, what is the average velocity of the flow? (*Ans.*: 0·15 ft, 0·26 fps)

1.7. An oil, of sp.gr. 0·92 and 50 cp viscosity, flows in a pipe of 6 in. diameter at a rate of 569 lb/min. What is the mass velocity, expressed in lb/sec×ft², the Reynolds number, the average velocity of the flow, in fps, and the velocity head, in feet, corresponding to this flow? (*Ans.*: 48·33, 716, 0·84, 0·022)

1.8. A jet of water strikes a stationary plate normally with a velocity of 30 fps. If the jet has a diameter of 1 in. what is the force, expressed in lb-f, exerted on the plate? (*Ans.*: 9·5)

1.9. Water is discharged horizontally from a stationary tank with a velocity of 20 fps. If the rate of discharge is 4 cfs, what is the reaction, expressed in lb-f, on the tank? (*Ans.*: 155)

1.10. A jet of water of 3-in. diameter impinges normally on a flat plate moving with its own velocity of 4 fps in the direction of the jet. If the rate of discharge is 1·2 cfs, what is the force, in lb-f, exerted on the plate, and the work done, in ft×lb-f/sec?

(*Ans.*: 39·8, 159·2)

1.11. A jet of water, of 2 in. diameter, strikes a flat plate inclined at an angle of 30° to the axis of the jet, with a velocity of 50 fps. What is the normal force, in lb-f, exerted on the plate, when the plate is: (a) stationary, (b) moving at 10 fps in the direction of the jet?

(*Ans.*: 52·8, 33·8)

1.12. Water is discharged through the nozzle of a fire-hose at the rate of 0·35 cfs. If the nozzle has a diameter of 1 in. and the velocity of water in the hose is 8 fps, what is the reaction of the jet, expressed in lb-f?   (*Ans.*: 38·13)

1.13. A jet issues vertically upwards from a nozzle with a velocity of 40 fps. What is the velocity of the jet at a height of 20 ft? At what height will the jet have zero velocity? In each case assume no losses, and no contraction of the jet. (*Ans.*: 17·66 fps, 24·84 ft).

1.14. Water is delivered through a section of a pipe line, which changes gradually from 6 in. to 12 in. diameter, at the rate of 18,000 lb/min. At the smaller diameter end of the section the pressure is 60 psia. What is the pressure at the larger diameter end if it is 20 ft higher than the other end of the section considered?   (*Ans.*: 55·12 psia)

1.15. A closed tank is partly filled with water, the air space above its surface being under pressure. Water is discharged from the tank through a hose of 2 in. diameter, the discharge end of which is open to the atmosphere 50 ft above the water surface in the tank. If the total loss in the flow is 18 ft water head, what must be the air pressure in the tank, expressed in psig, to discharge 0·45 cfs?   (*Ans.*: 32·36)

1.16. A gas, of density 0·076 lb/ft³, discharges from a tank to the atmosphere through an orifice of 0·5 in. diameter. When the pressure in the tank is 0·75 i.w.g. the rate of discharge is 0·215 lb/min. What is the coefficient of discharge at this instant?

(*Ans.*: 0·60)

1.17. A jet of water issues horizontally from an open tank through a small orifice, the centre of which is 3 ft below the water surface in the tank. If the jet is to be intercepted at a horizontal distance of 3 ft from the vena contracta, how many inches below the horizontal line passing through the centre of the orifice will the interception take place? Assume 0·98 for the coefficient of velocity.   (*Ans.*: 9·4)

1.18. Water is delivered by a duct at the rate of 8 cfs. The upper end of the duct has a 4 ft² cross-section and is 3 ft above the lower end, the cross-section of which is 1 ft². If the pressures at both ends are the same, what is the loss of head, expressed in feet of water, and in psi?   (*Ans.*: 2·07, 0·897)

1.19. In flow through a pipe of diameter (D), the pressure drop (Δp) along a distance of unit length of the pipe, may be related to the average velocity of flow (v) by the equation

$$\Delta p = k\,\frac{v^2\varrho}{D}\left[\frac{\mu}{Dv\varrho}\right]^t$$

in which (ϱ) and (μ) are the density and viscosity of the fluid, respectively, while (k) and (t) are some constants. How can this relationship be obtained by the application of dimensional analysis?

CHAPTER 2

FLOW MEASUREMENT IN PIPES

THE devices which are used for the measurement of flow of liquids and gases are called *flowmeters*. They consist of two essential elements having complementary functions. These are often referred to as the *primary* and *secondary* elements. In the primary element the process basic to measurement takes place while the secondary element is simply an indicating or recording instrument.

2.1. Classification of Flowmeters

The operating principle of the primary element forms the basis for a broad classification of flowmeters into two kinds, *quantity* and *inferential meters*. Quantity meters are essentially volumetric in operation, though in certain types of meters they may measure the flow by weighing. They are characterised by a cyclic displacement of one or more mechanical devices, having a reciprocating or rotary motion, and for this reason they are also known as *displacement meters*. Inferential meters do not measure the quantity directly. They measure the velocity of flow, and from this measurement infer the rate of flow.

All quantity meters, and certain types of inferential meters are classed as *mechanical meters*, the term mechanical indicating a mechanism in the path of the flow which moves continuously at a speed proportional to the rate of flow. The various types of mechanical meters which are available commercially are described in some detail in a number of books[5, 6] devoted to this subject.

The remaining types of inferential meters are non-mechanical in character. Their common feature is that they measure flow in terms of such quantities as kinetic energy, inertia, specific heat, or some other characteristic quantities. In an important class of meters of this type, the rate of flow is determined from measurement of the differential pressure created in the primary element of the meter. Since the successful operation of these meters largely depends on the accuracy in measuring the differential pressure and on its subsequent interpretation, this subject will be considered first in the following paragraphs.

2.2. Manometers

Devices which are used for the measurement of pressure difference between two points of a system are known as *manometers*. If one of the points is open to the atmosphere, the difference measured is the *gauge pressure* and the manometer measuring this difference is called a *pressure gauge*. If the measured pressure is that of the atmosphere, the manometer is called a *barometer*.

A simple manometer consists of a *U-tube* partly filled with a liquid, which is usually called the *manometric liquid*. If the manometric liquid is to measure the pressure of a liquid as well, the two liquids must be immiscible and the manometric liquid must be the heavier one. Mercury is the most common manometric liquid in use, and water, alcohol, carbon-tetrachloride and some light oils are alternatives.

Consider a U-tube manometer, the ends of which have been connected to two points of a system where the respective pressures are p and $p + \Delta p$, as shown in Fig. 2.1. Let h be the observed displacement of the manometric

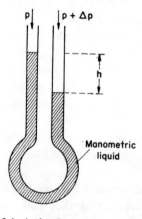

Fig. 2.1. A simple U-tube manometer.

liquid due to the pressure Δp in excess of p, then the relationship between Δp and h is obtained from a balance of pressure heads as follows.

Let γ_1 and γ be the densities of the manometric liquid and the fluid producing the differential effect respectively, then expressing the pressure heads in terms of the manometric liquid

$$\frac{p+\Delta p}{\gamma_1} + h\frac{\gamma}{\gamma_1} = \frac{p}{\gamma_1} + h$$

from which

$$\frac{\Delta p}{\gamma_1} = h\left(1 - \frac{\gamma}{\gamma_1}\right) \tag{2.1}$$

Similarly, expressing the pressure heads in terms of the lighter fluid of density γ

$$\frac{p+\Delta p}{\gamma} + h = \frac{p}{h} + h\frac{\gamma_1}{\gamma}$$

from which

$$\frac{\Delta p}{\gamma} = h\left(\frac{\gamma_1}{\gamma} - 1\right) \tag{2.2}$$

Let

$$\frac{\Delta p}{\gamma} = H$$

then

$$H = h\left(\frac{\gamma_1}{\gamma} - 1\right) \tag{2.3}$$

where H is the pressure head expressed in feet of the fluid the differential head of which is measured. If this fluid is a gas, its density (γ) will be considerably less than the density of the manometric liquid, and the ratio γ/γ_1 may often be ignored as negligibly small. Equation (2.1) then reduces to

$$\frac{\Delta p}{\gamma_1} = h \tag{2.4}$$

This may be written

$$\frac{\Delta p}{\gamma} = h\frac{\gamma_1}{\gamma}$$

or

$$H = h\frac{\gamma_1}{\gamma} \tag{2.5}$$

EXAMPLE 2.1

A U-tube contains mercury (sp.gr. = 13·6) as the manometric liquid. If one end of the tube is open to the atmosphere and the other connected to a water main, as shown in Fig. 2.2, the mercury displacement is 18 in. What is the gauge pressure in the main?

Solution

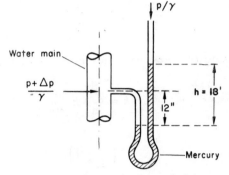

FIG. 2.2. Diagram for Example 2.1.

Let γ be the density of water, and Δp the gauge pressure in the main, then by a pressure head balance above the water–mercury interface in the tube

$$\frac{p+\Delta p}{\gamma}+\frac{12}{12} = \frac{18}{12}\left(\frac{13\cdot6}{1\cdot0}\right)+\frac{p}{\gamma}$$

from which

$$\frac{\Delta p}{\gamma} = (1\cdot5)\,(13\cdot6)-1\cdot0 = 19\cdot4 \text{ ft} \quad \text{(water)}$$

Taking $\gamma = 62\cdot4$ lb-wt/ft^3

$$\Delta p = (19\cdot4)\,(62\cdot4) = 1211\cdot6 \text{ psf}$$

This is equivalent to

$$\frac{1211\cdot6}{144} = 8\cdot41 \text{ psi}$$

EXAMPLE 2.2

A U-tube manometer is connected to two points of a vertical pipe
carrying an oil of sp.gr. 0·9, as shown in Fig. 2.3. If the manometric liquid
is water, and the two points are 2 ft apart, what is (a) the differential
pressure between the two points for a manometric reading of 6 in.? (b)
the manometer reading for a differential pressure of 1 psi?

Solution

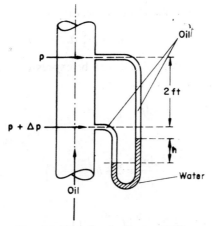

FIG. 2.3. Diagram for Example 2.2.

Let *h* be the manometric reading in each case, then balancing the pressure heads above the lower water–oil interface in the tube

$$\frac{p+\Delta p}{\gamma_1}+h\frac{\gamma}{\gamma_1} = \frac{p}{\gamma_1}+2\frac{\gamma}{\gamma_1}+h$$

from which

$$\frac{\Delta p}{\gamma_1} = (2-h)\frac{\gamma}{\gamma_1}+h \qquad\qquad (1)$$

where γ and γ_1 are the densities of the oil and water, respectively.
Since $\gamma/\gamma_1 = 0\cdot9$, then for case (a), using eq. (1)

$$\frac{\Delta p}{\gamma_1} = \left(2-\frac{6}{12}\right)(0\cdot9)+\frac{6}{12}$$

$$\frac{\Delta p}{\gamma_1} = 1\cdot85 \text{ ft} \qquad \text{(water)}$$

$$\Delta p = (1\cdot85)\,(62\cdot4) = 115\cdot44 \text{ psf}$$

This is equivalent to

$$\frac{115\cdot44}{144} = 0\cdot80 \text{ psi}$$

For case (b), since 1 psi is equivalent to $\Delta p = 144$ psf, then substituting this value in eq. (1) (taking $\gamma/\gamma_1 = 0\cdot9$)

$$\frac{144}{62\cdot4} = (2-h)(0\cdot9)+h$$

from which the required manometric reading is

$$h = 5\cdot1 \text{ ft}$$

Alternatively, the pressure heads can be expressed in terms of the lighter fluid, i.e. oil, as follows. For case (a)

$$\frac{p+\Delta p}{\gamma}+h = \frac{p}{\gamma}+2+h\frac{\gamma_1}{\gamma}$$

from which

$$\frac{\Delta p}{\gamma} = 2+h\left(\frac{\gamma_1}{\gamma}-1\right) \tag{2}$$

For $h = 6/12$ ft

$$\frac{\Delta p}{\gamma} = 2+\frac{6}{12}\left(\frac{1\cdot0}{0\cdot9}-1\right)$$

$$\frac{\Delta p}{\gamma} = \frac{1\cdot85}{0\cdot9} \text{ ft} \quad \text{(oil)}$$

$$\Delta p = \frac{1\cdot85}{0\cdot9}(0\cdot9)(62\cdot4) = 115\cdot44 \text{ psf}$$

(b) Substituting the data in eq. (2), and solving for h

$$\frac{144}{(0\cdot9)(62\cdot4)} = 2+h\left(\frac{1\cdot0}{0\cdot9}-1\right)$$

$$h = 5\cdot1 \text{ ft}$$

EXAMPLE 2.3

Two U-tubes, A and B, partly filled with liquids of different densities, are arranged as shown in Fig. 2.4. One end of A is connected to a pipe carrying a gas at a pressure of 1·7 psi below atmospheric pressure, while

one of the ends of B is open to the atmosphere. The other two ends are
connected by a common tube. The manometers read 1·5 ft in A and 1·25
ft in B. If the liquid in A has a sp.gr. of 1·6, what is the specific gravity of

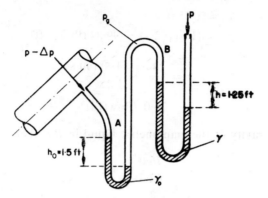

FIG. 2.4. Diagram for Example 2.3.

the liquid in B? The densities of the gas and the air trapped in the tubes
may be taken as negligibly small.

Let subscript o refer to the manometer A, the symbols referring to the
manometer B being left without qualification. Also let p_g be the common
pressure in the connecting tube, then the following balances can be set up.
For A

$$\frac{p-\Delta p}{\gamma_o}+h_o = \frac{p_g}{\gamma_o}$$

where p is the atmospheric pressure. From this equation

$$p_g = p-\Delta p + h_o\gamma_o \tag{1}$$

For B

$$\frac{p_g}{\gamma}+h = \frac{p}{\gamma}$$

from which

$$p_g = p-h\gamma \tag{2}$$

Equating (1) and (2)

$$p-h\gamma = p-\Delta p + h_o\gamma_o$$

from which

$$\gamma = \frac{\Delta p - h_o\gamma_o}{h}$$

Substituting in this equation for

$$\Delta p = (1 \cdot 7)\,(144) = 244 \cdot 8 \text{ psf}$$
$$h_o = 1 \cdot 5 \text{ ft}$$
$$h = 1 \cdot 25 \text{ ft}$$
$$\gamma_o = (1 \cdot 6)\,(62 \cdot 4) = 99 \cdot 84 \text{ (lb-wt)/ft}^3$$

results in

$$\gamma = \frac{244 \cdot 8 - 1 \cdot 5)\,(99 \cdot 84)}{1 \cdot 25}$$

$$\gamma = 76 \cdot 0 \text{ (lb-wt)/ft}^3$$

The specific gravity of the manometric liquid in B is then

$$\frac{76 \cdot 0}{62 \cdot 4} = 1 \cdot 22$$

2.3. Inverted U-tube

By inverting a U-tube, a type of manometer is obtained as shown in Fig. 2.5. Its advantage over the simple U-tube manometer is that it does not require any manometric liquid.

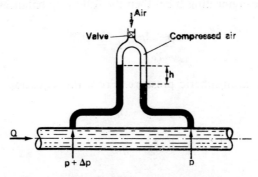

Fig. 2.5. The inverted U-tube manometer.

The bent part of the tube is usually provided with an air outlet, which is often fitted with a non-return valve. Through this valve, the air may be compressed into the tube, using a bicycle pump, in order that the menisci of the liquid can be made visible. If the compression is reasonably low, the density of the air above the menisci will be small relative to the density of the liquid, and the relationship between Δp and h may be presented approximately by

$$h = \frac{\Delta p}{\gamma} \qquad (2.6)$$

where γ is the density of the liquid.

EXAMPLE 2.4

An inverted U-tube is used to measure the differential pressure between two points in a pipe carrying an oil of 55·6 lb/ft³ density. If the density of the compressed air is 0·4 lb/ft³, what is the differential pressure for a reading of 10 in.?

Solution

With reference to Fig. 2.5, let γ_a be the density of the compressed air, then using eq. (2.1)

$$\frac{\Delta p}{\gamma} = h \left(1 - \frac{\gamma_a}{\gamma} \right)$$

in which γ is the density of the oil. Substituting the data in the above equation

$$\frac{\Delta p}{55·6} = \frac{10}{12} \left(1 - \frac{0·4}{55·6} \right)$$

from which

$$\Delta p = 44·0 \text{ psf} \quad (0·306 \text{ psi})$$

Should the term γ_a/γ be ignored, and eq. (2.6) used instead, the differential pressure would be

$$\Delta p = h\gamma = \frac{10}{12} (55·6)$$

$$\Delta p = 44·3 \text{ psf}$$

The error in using the approximate equation (2.6) in this example is less than 1·5 per cent, and this is acceptable in most engineering problems.

2.4. The Draught-gauge

If one limb of a U-tube is enlarged into a small reservoir and the other inclined, a kind of manometer is obtained as shown in Fig. 2.6. This kind of manometer is called the *draught-gauge*. It is commonly used for the measurement of small differential pressure, such as produced by furnace draught, from which it derives its name.

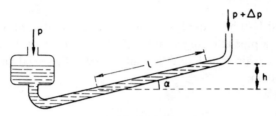

FIG. 2.6. The draught-gauge.

The reservoir is large enough to accommodate the manometric liquid displaced from the inclined limb without producing any noticeable change in it, so that the reading l is obtained directly on the scale along the incline. The vertical distance (h) is then calculated from the equation

$$h = l \sin \alpha$$

in which α is the angle of the inclination.

Most commonly, the angle is made such that $h/l = 1/10$, or the scale reading gives the vertical displacement of the manometric liquid directly in inches of water.

2.5. Venturi Meter

The flow in closed conduits can be conveniently derived from measurement of differential pressures produced in the primary elements of meters. One of the important devices making use of this principle is the *Venturi* meter. It consists of a tube, called the *Venturi tube*, in which the differential effect is created, and a manometer which registers the differential pressure. The tube is incorporated into the line carrying the fluid the flow of which is measured, so that it forms part of the line itself. A horizontal arrangement of a Venturi meter is shown in Fig. 2.7.

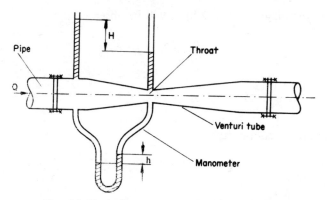

FIG. 2.7. Typical arrangement of a Venturi meter.

The tube narrows from both ends inward to a minimum, the narrowest section being called the throat. The differential head (H) is measured between the throat and a point upstream where the section is of the same diameter as that of the pipe to which the tube is flanged. By measurement of the differential head the flow is calculated from equations which can be derived as shown in Fig. 2.8.

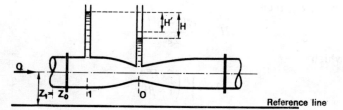

FIG. 2.8. The Venturi tube arranged horizontally.

Consider in Fig. 2.8 two sections, *1* and *0*, of a Venturi tube in a horizontal arrangement. Let us first assume that the fluid flowing through the tube behaves ideally, then applying Bernoulli's theorem between the two sections considered

$$Z_1 + \frac{p_1}{\gamma} + \frac{v_1^2}{2g} = Z_0 + \frac{p_0}{\gamma} + \frac{v_0^2}{2g} \qquad (2.7)$$

But for a horizontal arrangement $Z_1 = Z_0$, and

$$\frac{p_1 - p_0}{\gamma} = \frac{v_0^2 - v_1^2}{2g} \qquad (2.8)$$

Also from the diagram (Fig. 2.8), it will be observed that

$$\frac{p_1 - p_0}{\gamma} = H'$$

then

$$H' = \frac{v_0^2 - v_1^2}{2g} \tag{2.9}$$

From the law of continuity

$$Q = v_1 D_1^2 \pi/4 = v_0 D_0^2 \pi/4$$

from which

$$v_1 = v_0 \left(\frac{D_0}{D_1}\right)^2 \tag{2.10}$$

where D_1 is the diameter of the pipe carrying the fluid, assumed to be the same as the diameter of the tube at the section 1.

Substituting for v_1 in eq. (2.9)

$$H' = \frac{v_0^2[1 - (D_0/D_1)^4]}{2g}$$

Let $\beta = D_o/D_1$, then

$$H' = \frac{v_0^2[1 - \beta^4]}{2g} \tag{2.11}$$

from which

$$v_0 = \sqrt{\left(\frac{2gH'}{1 - \beta^4}\right)} \tag{2.12}$$

Now, let Q_0 be the theoretical flow referred to the section 0, then

$$Q_0 = A_0 v_0 = A_0 \sqrt{\left(\frac{2gH'}{1 - \beta^4}\right)} \tag{2.13}$$

Due to the energy loss in the flow of a real fluid, the observed differential head is greater then the theoretical one, and the actual velocity at the throat is less than that which could be calculated from eq. (2.12). Let H be the observed differential head, then the actual flow can be obtained from the equation

$$Q = C_D A_0 \sqrt{\left(\frac{2gH}{1 - \beta^4}\right)} \tag{2.14}$$

where C_D is the coefficient of discharge and Q is the actual volumetric rate of flow.

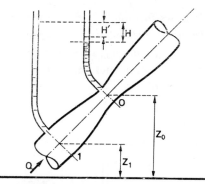

FIG. 2.9. The Venturi tube inclined to the horizontal.

For a Venturi tube inclined to the horizontal, eq. (2.14) can be modified as follows. From eq. (2.7), with reference to Fig. 2.9,

$$\frac{p_1 - p_0}{\gamma} = \frac{v_0^2 - v_1^2}{2g} + (Z_0 - Z_1) \tag{2.15}$$

Since

$$H' = \frac{p_1 - p_0}{\gamma}$$

and

$$\frac{v_0^2 - v_1^2}{2g} = \frac{v_0^2(1 - \beta^4)}{2g}$$

then

$$H' = \frac{v_0^2(1 - \beta^4)}{2g} + (Z_0 - Z_1) \tag{2.16}$$

from which

$$v_0 = \sqrt{\left(\frac{2g[H' + (Z_1 - Z_0)]}{1 - \beta^4}\right)} \tag{2.17}$$

and

$$Q_0 = A_0 v_0 = A_0 \sqrt{\left(\frac{2g[H' + (Z_1 - Z_0)]}{1 - \beta^4}\right)} \tag{2.18}$$

The actual flow is obtained by replacing the theoretical differential head (H') by the observed one (H) and by introducing the coefficient of discharge (C_D). The resulting equation is

$$Q = C_D A_0 \sqrt{\left(\frac{2g[H + (Z_1 - Z_0)]}{1 - \beta^4}\right)} \tag{2.19}$$

EXAMPLE 2.5

An oil (sp.gr. = 0·92) flows through a vertical tube of 8 in. diameter. The flow is measured by a Venturi tube of 4 in. diameter throat with a U-tube manometer, containing mercury, as shown in Fig. 2.10. If $C_D = 0·98$, what is (a) the flow, for a manometer reading of 9 in.? (b) the manometer reading, for a flow of 2 ft³/sec?

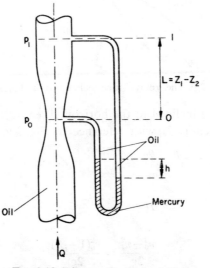

FIG. 2.10. Diagram for Example 2.5.

Solution

With reference to Fig. 2.10, eq. (2.16) becomes

$$H' = \frac{v_0^2(1-\beta^4)}{2g} - (Z_1 - Z_0)$$

Also by a balance of pressure heads above the lower meniscus of the mercury in the U-tube

$$\frac{p_0}{\gamma} + h\frac{\gamma_1}{\gamma} = \frac{p_1}{\gamma} - (Z_1 - Z_0) + h$$

from which

$$\frac{p_1 - p_0}{\gamma} = h\left(\frac{\gamma_1}{\gamma} - 1\right) - (Z_1 - Z_0)$$

But

$$H' = \frac{p_1 - p_0}{\gamma}$$

then

$$h\left(\frac{\gamma_1}{\gamma} - 1\right) + (Z_1 - Z_0) = \frac{v_0^2(1 - \beta^4)}{2g} - (Z_1 - Z_0)$$

from which

$$v_0 = \sqrt{\left(\frac{2gh(\gamma_1/\gamma - 1)}{1 - \beta^4}\right)}$$

The actual rate of flow is then given by

$$Q = C_D A_0 \sqrt{\left(\frac{2gh(\gamma_1/\gamma - 1)}{1 - \beta^4}\right)}$$

But $\qquad 1 - \beta^4 = 1 - (4/8)^4 = 15/16, \qquad$ then

(a) $\qquad h(\gamma_1/\gamma - 1) = (9/12)(13\cdot6/0\cdot92 - 1) = 10\cdot35$ ft

$$Q = (0\cdot98)\left[\left(\frac{4}{12}\right)^2 \frac{\pi}{4}\right]\sqrt{\left(\frac{(64\cdot4)(10\cdot35)}{15/16}\right)}$$

$$Q = 2\cdot278 \text{ cfs } (7848 \text{ lb/min})$$

(b) $\qquad 2 = (0\cdot98)\left[\left(\frac{4}{12}\right)^2 \frac{\pi}{4}\right]\sqrt{\left(\frac{64\cdot4h(13\cdot6/0\cdot92 - 1)}{15/16}\right)}$

from which

$$h = 0\cdot5965 \text{ ft } (7\cdot16 \text{ in.})$$

2.6. Orifice and Nozzle Meters

In its most familiar form, the primary element of an orifice meter consists of a thin plate, with a circular hole, clamped between a pair of flanges in a pipe line. The secondary element is a manometer connected to two taps located in the wall of the pipe, the downstream tap usually in the plane at the vena contracta.

The nozzle type of meter is virtually a Venturi tube without the downstream cone, fitted into a pipe like the orifice plate. The two types of meters are shown diagrammatically in Fig. 2.11.

The working principle of the orifice meter and nozzle is the same as that of the Venturi meter, and the same theory applies to all of them.

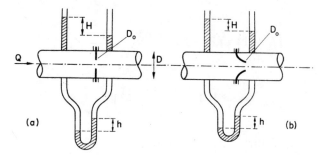

FIG. 2.11. The orifice meter (a), and the nozzle meter (b), in a horizontal arrangement.

EXAMPLE 2.6

The flow of water is measured by an orifice of 1 in. diameter inserted in a horizontal pipe of 1·61 in. diameter. The differential head is measured by a U-tube manometer containing mercury as the manometric liquid. If the coefficient of discharge is 0·61, what is the manometer reading for a flow of 150 lb/min?

Solution

Referring to Fig. 2.11(a)

$$\beta = D_o/D = 1\cdot00/1\cdot61$$
$$1 - \beta^4 = 1 - 0\cdot144 = 0\cdot856$$
$$A_o = D_o^2\pi/4 = (1/12)^2\pi/4 = 0\cdot00\,544 \text{ ft}^2$$
$$Q = \frac{150}{(62\cdot4)\,(60)} = 0\cdot0400 \text{ cfs}$$

Substituting the data in eq. (2.14)

$$0\cdot04 = (0\cdot61)\,(0\cdot00544)\sqrt{\left(\frac{64\cdot4H}{0\cdot856}\right)}$$

from which

$$H = 1\cdot932 \text{ ft}$$

Using eq. (2.3)

$$H = h \left(\frac{\gamma_1}{\gamma} - 1 \right)$$

The mercury–water density ratio is $\gamma_1/\gamma = 13\cdot6/1\cdot0 = 13\cdot6$, then

$$h = \frac{1\cdot932}{13\cdot6 - 1} = 0\cdot153 \text{ ft} \quad (1\cdot84 \text{ in.})$$

EXAMPLE 2.7

A gas-holder has developed a leak through a hole equivalent to an orifice of $\frac{1}{2}$ in. diameter. What is the rate of leak, if the pressure in the holder is 2 i.w.g.? A gas density of $0\cdot08$ lb/ft^3, and a coefficient of discharge of $0\cdot6$ may be assumed.

Solution

$$A_o = \left(\frac{1}{24} \right)^2 \frac{\pi}{4} = 0\cdot00136 \text{ ft}^2$$

$$1 - \beta^4 = 1$$

2 i.w.g. correspond to

$$H = \left(\frac{2}{12} \right) \left(\frac{62\cdot4}{0\cdot08} \right)$$

$$H = 130 \text{ ft (gas)}$$

Substituting in eq. (2.14)

$$Q = (0\cdot6)(0\cdot00136) \sqrt{\{(64\cdot4)(130)\}}$$
$$Q = 0\cdot0747 \text{ cfs}$$

This is equivalent to

$$(0\cdot0747)(0\cdot08)(60) = 0\cdot358 \text{ lb/min}$$

EXAMPLE 2.8

A nozzle is to be fitted into a horizontal pipe of 6 in. diameter, carrying an oil of specific gravity $0\cdot8$, for the purpose of flow measurement. The differential head is to be indicated by a U-tube manometer containing mercury. If the reading is not to exceed 12 in., when the flow is 40·8 lb/sec, what should be the diameter of the nozzle? A coefficient of discharge of 0·8 may be assumed.

Solution

With reference to Fig. 2.11(b)

$$\beta^4 = \left(\frac{D_o}{D}\right)^4 = \left(\frac{D_o}{0 \cdot 5}\right)^4 = 16D_o^4$$

$$A_o = D_o^2\,\frac{\pi}{4}$$

$$Q = \frac{40 \cdot 8}{(0 \cdot 8)\,(62 \cdot 4)} = 0 \cdot 817 \text{ cfs}$$

Using eq. (2.2)

$$H = h\left(\frac{\gamma_1}{\gamma}-1\right) = \left(\frac{12}{12}\right)\left(\frac{13 \cdot 6}{0 \cdot 8}-1\right)$$

$$H = 16 \text{ ft} \quad (\text{oil})$$

Substituting the data in eq. (2.14)

$$0 \cdot 817 = (0 \cdot 8)\left(D_o^2\,\frac{\pi}{4}\right)\sqrt{\left(\frac{(64 \cdot 4)\,(16)}{1-16D_o^4}\right)}$$

$$\sqrt{\left(\frac{1-16D_o^4}{D_o^4}\right)} = \frac{0 \cdot 8\pi\,\sqrt{\{(64 \cdot 4)\,(16)\}}}{(4)\,(0 \cdot 817)}$$

$$\sqrt{\left(\frac{1}{D_o^4}-16\right)} = 24 \cdot 68$$

$$\frac{1}{D_o^4}-16 = 609$$

$$\frac{1}{D_o^4} = 625$$

$$D_o = 0 \cdot 20 \text{ ft} \quad (2 \cdot 4 \text{ in.})$$

2.7. The Pitot Tube

The Pitot tube is a simple device used for the determination of point velocities in flowing fluids. In its simplest form, it consists of two parallel tubes mounted vertically, one of the tubes being straight while the other is bent at the base. When the tubes are immersed in a liquid, with the bent end facing upstream, the liquid in the bent tube will rise to a height H above that in the straight tube, as shown in Fig. 2.12.

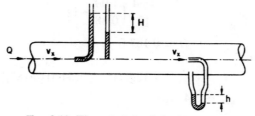

FIG. 2.12. The principle of the Pitot tube.

The differential effect in the Pitot tube is due to the velocity of the fluid, i.e. its kinetic energy, at the point of measurement. Let this point velocity be v_x, then the differential head produced is given by

$$H = \frac{v_x^2}{2g} \tag{2.20}$$

The velocity head is usually measured by a manometer, and this makes the Pitot tube a practical instrument for the measurement of point velocities in gases as well as liquids.

From a series of measurements across a stream, the average velocity can be evaluated by a process of integration. This is often a laborious procedure, and in case of circular ducts, such as pipes, it can be simplified by taking only a single measurement at their axis, where the point velocity is a maximum ($v_x = v_{max}$), and calculating the average velocity from the relationship

$$v = \alpha v_{max} \tag{2.21}$$

in which the factor α may be obtained from Fig. 1.7, upon the evaluation of the Reynolds number.

EXAMPLE 2.9

An oil (sp.gr. = 0·88, viscosity = 50 cp) flows in a pipe of 3 in. diameter. The flow is measured by a Pitot tube located centrally, and the differential head is indicated by a U-tube, containing water as the manometric liquid. What is the rate of flow for a reading of 16·4 in.?

Solution

Using eq. (2.3)

$$H = \frac{16\cdot4}{12}\left(\frac{1\cdot00}{0\cdot88}-1\right)$$

$$H = 0\cdot187 \text{ ft}$$

From eq. (2.20)

$$v_x = v_{max} = \sqrt{(2gH)} = \sqrt{\{(64\cdot4)(0\cdot187)\}}$$

$$v_{max} = 3\cdot46 \text{ fps}$$

Assuming laminar flow, $\alpha = 0\cdot5$, and (from eq. (2.21) the average velocity is

$$v = 1\cdot73 \text{ fps}$$

$$Q = vA = (1\cdot73)\left(\frac{3}{12}\right)^2\left(\frac{\pi}{4}\right)$$

$$Q = 0\cdot085 \text{ cfs}$$

This is equivalent to 5·1 cfm, or to

$$(5\cdot1)\,(0\cdot88)\,(62\cdot4) = 280 \text{ lb/min}$$

Checking on the Reynolds number

$$Re = \frac{Dv\varrho}{\mu} = \frac{(3/12)\,(1\cdot73)\,(0\cdot88)\,(62\cdot4)}{(50)\,(0\cdot000672)}$$

$$Re = 707$$

This is less than the critical number 2000, thus proving the correctness of the assumption made.

2.8. Area Meters

The Venturi tube, orifice and nozzle meters, as well as the Pitot tube, are based on a common principle, namely that they measure the flow in terms of the differential head created in their primary elements. For this reason they are classed as *differential head meters*. For any of these meters, the flow can be calculated from a general equation of the form

$$Q = \text{constant }\sqrt{H}$$

where H is the differential head produced in the primary element of the meter.

A basically different principle is utilised in a class of meters known as *area meters*. The differential pressure head in this class of meters remains constant within the whole range of flow, but instead the flow area changes with the flow according to the equation

$$Q = \text{constant } (A_o)$$

in which A_o is the flow area of the meter.

The most popular meter in this class consists of a gently tapered tube, mounted vertically, with the bore increasing upwards. The tube encloses a float which is only slightly smaller than the bore of the tube. The flow is determined from the position of the float, which is indicated on a scale calibrated in any convenient units.

The float is usually shaped in the form of a top, with its rim grooved to produce a rotary movement for the purpose of stability. This kind of motion gives rise to a name *rotameter* under which this type of area meter is commonly known. The theory pertinent to rotameters may be underlined with reference to Fig. 2.13, as follows.

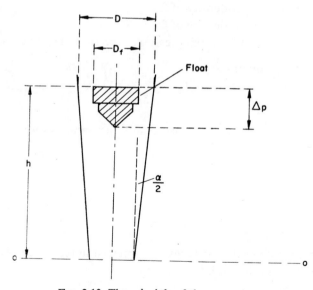

FIG. 2.13. The principle of the area meter.

Consider the position of the float when it is at a vertical distance h from the lowest mark on the scale (level 0–0 in Fig. 2.13). Let Q be the flow corresponding to this position, then ignoring the $(1-\beta^4)$ term of eq. (2.14), as negligibly small in this case,

$$Q = C_D A_o \sqrt{(2gH)} \tag{2.22}$$

Let D_f be the diameter of the float and D the variable bore of the tube, then the flow area

$$A_o = (D^2 - D_f^2)\pi/4 \tag{2.23}$$

Since the differential head remains unchanged at all positions of the float, it follows that

$$H = \frac{\Delta p}{\gamma} = \text{constant}$$

The float remains at equilibrium when the force acting on its base equals the weight of the float. Since the base has an area of $D_f^2\pi/4$, then from the balance of forces (allowing for the buoyancy effect)

$$(\Delta p)D_f^2\pi/4 = V_f(\gamma_f - \gamma) \tag{2.24}$$

where V_f = volume of float,
           γ_f = weight-density of float,
           γ = weight-density of fluid,
     (Δp) = the differential pressure, expressed in engineering units.

From the last two equations

$$H = \frac{\Delta p}{\gamma} = \frac{V_f(\gamma_f - \gamma)}{\gamma D_f^2\pi/4}$$

The ratios of weight and mass densities are the same, so we may write

$$\frac{\gamma_f - \gamma}{\gamma} = \frac{\varrho_f - \varrho}{\varrho}$$

and

$$H = \frac{V_f(\varrho_f - \varrho)}{\varrho D_f^2\pi/4} \tag{2.25}$$

The variable bore of the tube may be expressed in terms of the diameter of the float (D_f) as follows. Let

$$D = D_f + \sigma D$$

where σD is the variable clearance expressed in terms of D between the base of the float and the inside wall of the tube. It follows that

$$D^2 - D_f^2 = (D_f + \sigma D)^2 - D_f^2 = 2D_f(\sigma D) + (\sigma D)^2$$

Ignoring the second-order term $(\sigma D)^2$ of this equation as negligibly small,

$$D^2 - D_f^2 = 2D_f(\sigma D) \tag{2.26}$$

Let α be the cone angle, then with reference to Fig. 2.13

$$\frac{\sigma D}{2} = h \tan \frac{\alpha}{2}$$

Substituting from this equation for (σD) in eq. (2.26)

$$D^2 - D_f^2 = 4hD_f \tan \frac{\alpha}{2}$$

Again, substituting from this equation in eq. (2.23), and simplifying

$$A_o = hD_f\pi \tan \frac{\alpha}{2}$$

Finally, substituting for A_o from this equation, and for H from eq. (2.25), in eq. (2.22)

$$Q = C_D hD_f\pi \tan \frac{\alpha}{2} \sqrt{\left(2g\frac{V_f(\varrho_f - \varrho)}{\varrho D_f^2\pi/4}\right)}$$

Simplifying

$$Q = C_D h \tan \frac{\alpha}{2} \sqrt{(8\pi g V_f)} \sqrt{\frac{\varrho_f - \varrho}{\varrho}} \tag{2.27}$$

For a given meter, the constants can be grouped together, and eq. (2.27) presented in the form

$$Q = K \sqrt{\left(\frac{\varrho_f - \varrho}{\varrho}\right)} \tag{2.28}$$

where K is a constant.

EXAMPLE 2.10

A rotameter, calibrated for metering a liquid of density 62 lb/ft³, has a scale ranging from 0·05 to 0·5 cfm. It is intended to use this meter for metering a gas, of density 0·08 lb/ft³ within the flow range between 1·0 and 10 cfm. What should be the density of the new float, if the original one has a density of 120 lb/ft³? Both floats are assumed to have the same volume and shape.

Solution

Let subscript (*o*) refer to the original set, then from eq. (2.28)

$$Q_o = K \sqrt{\left(\frac{\varrho_{fo} - \varrho_o}{\varrho_o} \right)} \qquad (1)$$

Similarly for the gas

$$Q = K \sqrt{\left(\frac{\varrho_f - \varrho}{\varrho} \right)} \qquad (2)$$

where ϱ_f is the required density of the new float. The scale ratio is

$$\frac{1 \cdot 00}{0 \cdot 05} = \frac{10}{0 \cdot 5} = 20$$

It follows that

$$Q = 20 Q_o$$

Substituting for Q in eq. (2)

$$20 Q_o = K \sqrt{\left(\frac{\varrho_f - \varrho}{\varrho} \right)} \qquad (3)$$

From the equations (1) and (3)

$$20 = \sqrt{\left(\frac{(\varrho_f - \varrho) \varrho_o}{(\varrho_{fo} - \varrho_o) \varrho} \right)}$$

Substituting the data

$$20 = \sqrt{\left(\frac{(\varrho_f - 0 \cdot 08) \, (62)}{(120 - 62) \, (0 \cdot 08)} \right)}$$

from which

$$\varrho_f = 30 \cdot 0 \ \text{lb/ft}^3$$

2.9. Thermal and Dilution Meters

If heat is supplied at a steady rate to one section of a flow system, then from the resulting rise in temperature the flow can be determined from a simple heat balance. Alternatively, this can be achieved by analysing the cooling effect of a flowing fluid on a hot body immersed in its stream.

Similarly, the flow can be determined by introducing into a flow system a foreign substance, the concentration of which can be readily determined in the metered fluid. Flow of water, for example, can be measured very accurately by injecting common salt at a steady rate at one section of water stream, and by analysing the water downstream where the salt is supposed to have dissolved and mixed thoroughly with the water.

The techniques which make use of the above method of flow measurement give rise to the name of *thermal* or *dilution meters*.

EXAMPLE 2.11

The flow of a gas, of specific heat = 0·24 Btu/(lb) (°F), is measured by supplying heat through an electrical resistance wire and recording the temperature rise of the gas. In a test, when a current of 2 Amp. was passed through the wire, the observed temperature rise was 6°F. Assuming a 10 per cent loss in heat, what is the rate of flow of the gas?

Solution
$$1 \text{ kW} = 1·341 \text{ hp} = 1·341 \times 550 \text{ (lb-f) (ft/sec)}$$

Since 1 Btu = 778 (lb-f) (ft), then

$$1 \text{ kW} = \frac{(1·341)(550)}{778} = 0·948 \text{ Btu/sec}$$

At a loss of 10 per cent, the rate of heat supply is

$$(2)(0·9)(0·948) \text{ Btu/sec}$$

By a heat balance

$$\text{Flow rate (lb/sec)} = \frac{(2)(0·9)(0·948)}{(0·24)(6)}$$

$$\text{Flow rate} = 1·185 \text{ lb/sec}$$

EXAMPLE 2.12

A steam meter was calibrated by a dilution method, using anhydrous ammonia. In a test, 6 lb of ammonia were blown into the steam main for 5 min at a steady rate. Further downstream, a sample of the steam was condensed for titration; 15·5 ml of N/1 acid were required to neutralise ammonia in 100 gm of the condensate. At the same time the meter reading was 44·7 lb/min. What percentage error does this figure represent, if the dilution method is assumed to give a correct answer?

Solution

Ammonia, rate of feed $= \frac{6}{5} = 1\cdot2$ lb/min.

Taking 17 for the molecular weight of ammonia, the ammonia content in 100 gm of the condensed steam is

$$(15\cdot5)\,\frac{17}{1000} = 0\cdot264 \text{ gm}$$

The rate of steam flow is therefore

$$\frac{100-0\cdot264}{0\cdot264}\,(1\cdot2) = 45\cdot3 \text{ lb/min}$$

This represents an error of

$$45\cdot3-44\cdot7 = 0\cdot6 \text{ lb/min}$$

On a percentage basis, the error is

$$\frac{0\cdot6}{45\cdot3}\,(100) = 1\cdot325 \text{ per cent}$$

EXAMPLE 2.13

The flow of dry chlorine, containing $1\cdot4$ mole per cent oxygen (average mol.wt. $= 70\cdot5$), is metered in a pipe of 3 in. diameter, by a nozzle of $1\frac{1}{2}$ in. diameter, and U-tube, with an oil of density 53 lb/ft as a manometric liquid. For the purpose of calibrating the nozzle meter, oxygen (mol.wt. $= 32$) is blown into one point of the pipe from a weighed cylinder.

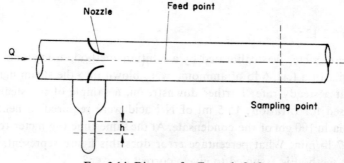

FIG. 2.14. Diagram for Example 2.13.

In one run, the cylinder loses 2·2 lb in weight in 5 min, and a sample taken downstream from the feed point shows 8·4 mole per cent of oxygen. During the run, the manometer reading was 9 in. If the density of the chlorine gas is 0·172 lb/ft³, what is the coefficient of discharge for the nozzle meter?

Solution

Rate of feed of oxygen = $2·2/[(32)(5)] = 0·01375$ lb-mole/min. Let x and y be the mole-rates of flow of the gas upstream and downstream of the feed point respectively, then by an overall material balance

$$x + 0·01375 = y$$

and by an oxygen balance

$$0·014x + 0·01375 = 0·084y$$

Solving simultaneously

$$x = 0·18 \text{ lb-mole/min}$$

or

$$\frac{(0·18)(70·5)}{60} = 0·212 \text{ lb/sec}$$

For the nozzle meter

$$1 - \beta^4 = 1 - \left(\frac{1·5}{3·0}\right)^4 = 0·9375$$

$$A_o = \left(\frac{1·5}{12}\right)^2 \frac{\pi}{4} = 0·01225 \text{ ft}^2$$

From eq. (2.3)

$$H = \frac{9}{12}\left(\frac{53·0}{0·172} - 1\right)$$

$$H = 230 \text{ ft}$$

$$Q = \frac{0·212}{0·172} \quad \text{(cfs)}$$

Substituting in eq. (2.14)

$$\frac{0·212}{0·172} = C_D(0·01225)\sqrt{\left(\frac{(64·4)(230)}{0·9375}\right)}$$

from which

$$C_D = 0·80$$

Problems

2.1. A U-tube manometer measures the pressure of a gas, using water as a manometric liquid. If the density of the gas is negligible, what is its pressure when the displacement of water is 3 in.? (*Ans.*: 0·11 psi)

2.2. A U-tube manometer, containing mercury (sp.gr. = 13·6) as the manometric liquid, is connected to two points of a horizontal pipe carrying a solution of sp.gr. 1·2. What is the displacement, in inches, for a differential pressure of 0·67 psi? Both limbs of the tube contain the solution above the mercury levels. (*Ans.*: 1·5 in)

2.3. A U-tube manometer is connected to two points of a horizontal pipe carrying an oil of sp.gr. 0·92. The manometric liquid is mercury. What is the differential pressure for a reading of 8 in.? (*Ans.*: 3·66 psi)

2.4. A gas (density = 0·0243 lb/ft³) flows through a pipe of 3 in. diameter. The flow is measured by a Venturi tube of 1·5 in. diameter throat and a U-tube manometer containing mercury. What is the flow for a manometer reading of 8 in.? The coefficient of discharge is 0·96. (*Ans.*: 21·7 lb/min)

2.5. The flow of a gas (density = 0·08 lb/ft³) is measured by an orifice meter inserted in a 4-in. diameter pipe. The diameter of the orifice is 2 in., and a U-tube containing water is used as a manometer. What is the flow of the gas for a manometer reading of 3 in.? The coefficient of discharge C_D = 0·6. (*Ans.*: 7·27 lb/min)

2.6. A pipe, of 6 in. diameter, carries a solution (sp.gr. = 1·2) at a rate of 84 lb/min, as measured by a horizontal Venturi and a U-tube manometer containing mercury. If, for this flow, the manometer reads 12 in., what is the diameter of the throat of the Venturi tube, assuming C_D = 0·97, and ignoring the coefficient of approach, i.e. the β term of eq. (2.14)? (*Ans.*: 2·88 in.)

2.7. Water flows in a pipe of 12 in. diameter. Its flow is measured by a horizontal Venturi tube of 4 in. diameter throat and a U-tube manometer containing mercury as the manometric liquid. What is the rate of flow for a manometer reading of 2·2 in., if C_D = 0·98? (*Ans.*: 1·05 cfs)

2.8. A Venturi meter, of 6 in. diameter throat, is installed horizontally in a pipe of 12 in. diameter. A U-tube manometer contains a liquid of an unknown density. If the manometer reading is 2·08 ft when water flows at a rate of 2·0 cfs, what is the specific gravity of the manometric liquid, given C_D = 0·985? (*Ans.*: 1·75)

2.9. A Venturi meter, of 2 in. diameter throat, is installed in a vertical pipe of 4 in. diameter. The pipe carries an oil (sp.gr. = 0·9), and pressure gauges are connected to the throat and to a point 3 ft above it. When the flow is upwards, the two gauges read the same. Assuming C_D = 0·97, what is the rate of flow? (*Ans.*: 17·05 lb/sec)

2.10. The flow of a gas (density = 0·077 lb/ft³) is measured by a Pitot tube located centrally in a pipe of 6·065 in. diameter. The reading on an inclined manometer (inclination = 10 : 1), using water as a manometric liquid, is 1 in. If the average velocity is 0·81 of the maximum velocity at the centre of the pipe, what is the flow? (*Ans.*: 15·8 lb/min)

2.11. A Pitot tube measures the flow of an oil (sp.gr. = 0·92) carried in a pipe of 4·03 in. diameter. The tube is located centrally in the pipe, and a U-tube manometer used with the Pitot tube contains water as the manometric liquid. What is the rate of flow of the oil for a manometer reading of 21·2 in? The flow is laminar, and the average velocity is half the velocity at the centre of the pipe. (*Ans.*: 8·25 cfm)

2.12. A gas (density $= 0.0495$ lb/ft³) flows in a pipe of 9 in. diameter at a rate of 28,670 ft³/hr. The flow is measured by a Pitot tube located centrally in the pipe, and connected to an inclined manometer (inclination $= 10 : 1$). It contains water as the manometric liquid. If the average velocity of the gas is 0.8 maximum velocity, what is the manometer reading for this flow? (*Ans.*: 0.75 in.)

2.13. A rotameter, which has been calibrated in cfm for water, is to be used for metering a solution of sp.gr. of 1.25. For this purpose the density of the float has been increased from 150 lb/ft³ to 180 lb/ft³, without altering its volume and shape.

What correction factor should be introduced to the original scale, in order that it can be used for the solution? (*Ans.*: 0.965)

2.14. The flow of a gas, of specific heat 0.24 Btu/(lb)(°F), is measured by a thermal method, using a resistance wire for the supply of heat. In a test, a current of 0.88 Amp. was passed through the wire and the observed rise in temperature was 2.97°F. Assuming no heat losses, and taking 1 kW $= 0.948$ Btu/sec, what is the flow of the gas? (*Ans.*: 1.17 lb/sec)

2.15. A gas (mol.wt. $= 30$) contains 2 mole per cent of oxygen. Its flow is measured by a dilution method using oxygen from a cylinder. In one test, 64 lb of oxygen were blown into the gas stream for 1 min, and the analysis of the gas downstream gave 5 mole per cent of oxygen. If the temperature and pressure of the gas were 20°C and 16 psia, respectively, what was its volumetric rate of flow? (*Ans.*: 22,410 cfm)

CHAPTER 3

FLOW IN PIPES

WHEN a solid body is caused to slide over a dry surface of another solid body, resistance is experienced due to friction. This resistance is proportional to the normal force at the surface of contact but is practically independent of the area of contact and of the sliding velocity. Resistance is also experienced when a fluid is in relative motion with a solid surface. In this case, however, the resistance depends significantly on the area of contact and on the relative velocity. The difference between the two behaviours is so fundamental that no analogy can be drawn between them, except perhaps for the common heat effects, resulting from the dissipation of energy.

3.1. Basic Resistance Equation

The basic resistance equation in fluid mechanics may be deduced by means of dimensional analysis as follows:

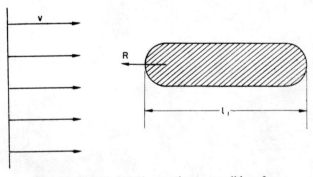

FIG. 3.1. Fluid in relative motion to a solid surface.

Consider, in Fig. 3.1, a fluid approaching a stationary body with a uniform velocity v. Let the characteristic dimension of the body be its length l, and let ϱ and μ be the density and viscosity of the fluid, respectively. Then, applying the principle of dimensional analysis, the following equation can be set up.

$$R = kl^x v^y \varrho^z \mu^t \qquad (3.1)$$

where R is the force resisting the motion, or that which is required to keep the body stationary, and k is a proportionality constant.

Substituting the dimensions for variables of eq. (3.1)

$$ML\theta^{-2} = (L)^x (L\theta^{-1})^y (ML^{-3})^z (ML^{-1}\theta^{-1})^t$$

Equating the indices for
Mass (M)

$$1 = z + t$$

Time (θ)

$$-2 = -y - t$$

Length (L)

$$1 = x + y - 3z - t$$

Solving the three equations simultaneously in terms of t

$$z = 1 - t$$
$$y = 2 - t$$
$$x = 2 - t$$

Substituting for x, y, and z in eq. (3.1)

$$R = k(l)^{2-t} (v)^{2-t} (\varrho)^{1-t} (\mu)^t$$

Rearranging the terms

$$R = kl^2 v^2 \varrho \left[\frac{\mu}{lv\varrho} \right]^t \qquad (3.2)$$

Let

$$C_D' = k \left[\frac{\mu}{lv\varrho} \right]^t$$

then

$$R = C_D' \varrho l^2 v^2 \qquad (3.3)$$

Now, let S be the surface area of the solid body. This area is some function of the leading dimension l, hence

$$S = k'l^2 \qquad (3.4)$$

where k' is another proportionality constant.

Substituting for l^2 from eq. (3.4) in eq. (3.3)

$$R = \frac{C_D'}{k'} S\varrho v^2$$

Let

$$C_D = \frac{C_D'}{k'}$$

then

$$R = C_D S\varrho v^2 \tag{3.5}$$

Also letting $R' = R/S$

$$R' = C_D \varrho v^2 \tag{3.6}$$

where C_D is called the *drag coefficient* or *drag factor*, and R' is the resistance per unit surface area, or per unit of projected area of the body on a plane normal to the direction of motion, depending on how the drag factor is defined.

Equation (3.6) is the basic resistance equation in fluid mechanics. A similar form was deduced by Newton through analysis of the actual mechanism of drag in flow around immersed bodies.

Only in the most elementary situations can the drag factor be determined analytically. In most practical cases, however, its determination is based on experimental results, although analytical methods are often helpful in systematising, explaining, and extrapolating experimental data.

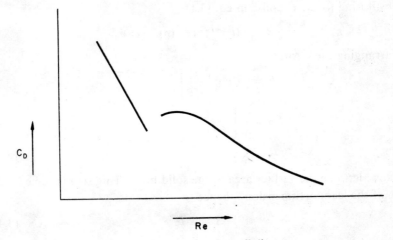

Fig. 3.2. The Stanton–Pannell diagram.

From the equations (3.2), (3.3) and (3.6), it is evident that the drag factor is some function of the group of terms $(\mu/lv\varrho)$, which is the reciprocal of the Reynolds number. This number has, therefore, become the most convenient parameter in use for the determination of the drag factor.

Blasius was the first to present the experimental data, partly from his own work, in the form of a plot of C_D vs. Re. Nevertheless, the diagram itself is best known as the Stanton–Pannell diagram (Fig. 3.2), in recognition of the contributions made, mainly by Stanton, in the field of pipe resistance and in many other aspects of experimental fluid mechanics.

The Stanton–Pannell diagram, which applies to flow in circular conduits, shows a straight line relationship between C_D and Re for the laminar range. From the slope of the line, the following relationship can be established:

$$C_D = \frac{8}{Re} \tag{3.7}$$

in which

$$Re = \frac{Dv\varrho}{\mu} = \frac{DG}{\mu} \tag{3.8}$$

The symbol D in this equation represents the pipe diameter, as a characteristic dimension.

EXAMPLE 3.1

Water (density $= 62\cdot4$ lb/ft^3, viscosity $= 0\cdot000672$ lb/(ft) (sec) flows in a tube of $0\cdot25$ in. bore, at a Reynolds number of 2000. What resistance is offered by a 10 ft length of the tube?

Solution

From eq. (3.8)

$$v = \frac{\mu Re}{D\varrho} = \frac{(0\cdot000672)\,(2000)}{(0\cdot25/12)\,(62\cdot4)}$$

$$v = 1\cdot034 \text{ fps}$$

At $Re = 2000$, the flow is still laminar, then from eq. (3.6) and eq. (3.7)

$$C_D = \frac{R'}{\varrho v^2} = \frac{8}{Re}$$

$$\frac{R'_{\;i}}{(62\cdot4)\,(1\cdot034)^2} = \frac{8}{2000}$$

$$R' = 0\cdot2657 \; \text{poundal/ft}^2$$

In 10 ft length, the area of the surface is

$$S = (0\cdot25/12)\,(10\pi) = 0\cdot654 \; \text{ft}^2$$
$$R = R'S = (0\cdot2657)\,(0\cdot654)$$

or $\qquad R = 0\cdot1738 \; \text{poundals}$

$$\frac{0\cdot1738}{32\cdot2} = 0\cdot0054 \; \text{lb-f}$$

3.2. The Darcy Equation

Consider, in Fig. 3.3, a fully developed flow in a straight pipe of dia-meter D. The cross-sectional area of the pipe is then $D^2 \pi/4$, and the sur-face area in a length L is πDL. Let $\Delta p = p_1 - p_2$ be the drop in pressure experienced in this length of the pipe, then by a balance of forces

$$R'\pi DL = (\Delta p)\, g_c(D^2\pi/4)$$

from which

$$R' = (\Delta p)g_c \frac{D}{4L} \tag{3.9}$$

where g_c is the gravitational conversion factor to allow for the fact that, unlike the resistance R, the pressure drop Δp is commonly expressed in the gravitational system of units.

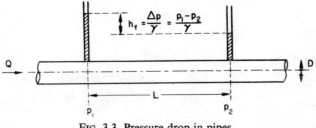

FIG. 3.3. Pressure drop in pipes.

Substituting from eq. (3.6), for R' in eq. (3.9)

$$C_D \varrho v^2 = (\Delta p) g_c \frac{D}{4L} \qquad (3.10)$$

from which

$$\frac{(\Delta p) g_c}{\varrho} = \frac{4 C_D v^2 L}{D} \qquad (3.11)$$

Substituting (from eq. (1.6)) for $\varrho = \gamma (g_c/g)$

$$\frac{(\Delta p) g_c}{\gamma (g_c/g)} = \frac{4 C_D v^2 L}{D}$$

Simplifying and rearranging the terms

$$\frac{\Delta p}{\gamma} = \frac{4 v^2 L C_D}{gD} \qquad (3.12)$$

Let

$$f = 2 C_D = 2 \frac{R'}{\varrho v^2} \qquad (3.13)$$

and

$$h_f = \frac{\Delta p}{\gamma} \qquad (3.14)$$

then

$$h_f = 4f \frac{L}{D} \left(\frac{v^2}{2g} \right) \qquad (3.15)$$

This equation is still the basic form used in engineering practice for the determination of the loss of head due to friction in straight pipes. In this form, it was first suggested by Weisbach (1806–71), and the equation bears his name in most European countries. In Great Britain, however, eq. (3.15) is commonly known as the *Darcy equation*. Under this name it is also known in the United States of America, although in some American texts eq. (3.15) is also called the *Fanning equation*. In this text, the name Darcy will be retained for the equation, and the factor f will be called *the Darcy friction factor*. This factor is normally obtained from a diagram, reproduced here in Fig. 3.4.

The Darcy friction diagram is a plot of f against Re on a logarithmic scale.

The two slightly curved lines, B and C, shown on the diagram, refer to very smooth surfaces, such as made by glass, copper or nickel, and to clean commercial steel pipes, respectively.

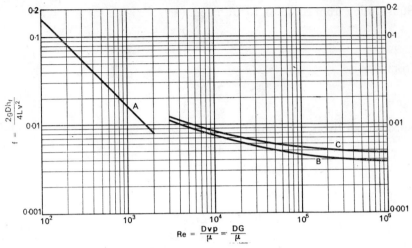

FIG. 3.4. The Darcy friction factor diagram.

Substituting in eq. (3.15) for

$$v = \frac{Q}{A} = \frac{4Q}{\pi D^2}$$

$$h_f = 4f \left(\frac{L}{D}\right) \left(\frac{1}{2g}\right) \left(\frac{16Q^2}{\pi^2 D^4}\right)$$

Rearranging and taking $2g = 64\cdot4$

$$h_f = \frac{(4)(16)}{(\pi^2)(64\cdot4)} \left[\frac{fLQ^2}{D^5}\right]$$

But, since

$$\frac{(\pi^2)(64\cdot4)}{(4)(16)} \cong 10$$

then

$$h_f = \frac{fLQ^2}{10D^5} \qquad (3.16)$$

From the slope of the straight line A, in eq. (3.4), which terminates at $Re = 2000$, approximately, the Darcy friction factor in laminar flow is given by

$$f = \frac{16}{Re} \qquad (3.17)$$

This is what one would naturally expect from the fact that $f = 2C_D$.

In the case of turbulent flow, the friction factor depends more on the roughness pattern of the pipe material than on the Reynolds number, and for this reason the factor cannot be simply presented in a mathematical form. For smooth pipes, however, Blasius has shown that for Reynolds numbers between 3000 and 100,000, the curve B in Fig. 3.4 may be closely approximated by a line whose equation is

$$f = 0.079/Re^{0.25} \tag{3.18}$$

When this is substituted in eq. (3.15), the loss of head h_f will be found to be proportional to $v^{1.75}$.

EXAMPLE 3.2

An oil (density = 59·3 lb/ft³, viscosity = 50 cp) flows through a horizontal pipe of 4 in. diameter. What is the drop in pressure, and the theoretical horse-power required in a mile length (5280 ft) of the pipe, if the oil flows at a rate of (a) 10 lb/sec, (b) 50 lb/sec?

It may be assumed that in turbulent flow, the Darcy friction factor is 10 per cent higher than calculated for a smooth pipe.

Solution

(a)
$$Q = \frac{10 \text{ lb/sec}}{59 \cdot 3 \text{ lb/ft}^3} = 0 \cdot 1686 \text{ ft}^3/\text{sec}$$

$$A = D^2 \pi/4 = (4/12)^2 \pi/4 = \pi/36 \text{ ft}^2$$

$$v = \frac{Q}{A} = \frac{0 \cdot 1686}{\pi/36} = 1 \cdot 932 \text{ fps}$$

$$Re = \frac{Dv\varrho}{\mu} = \frac{(4/12)(1 \cdot 932)(59 \cdot 3)}{(50)(0 \cdot 000672)}$$

$$Re = 1145$$

The Reynolds number is less than 2000, hence the flow is laminar and from eq. (3.17)

$$f = \frac{16}{Re} = \frac{16}{1145}$$

From eq. (3.15)

$$h_f = 4f \frac{L}{D} \frac{v^2}{2g} = (4) \left(\frac{16}{1145} \right) \left(\frac{5280}{4/12} \right) \left(\frac{1 \cdot 932^2}{64 \cdot 4} \right)$$

$$h_f = 51 \cdot 7 \text{ ft}$$

From eq. (3.14)

$$\Delta p = \gamma h_f = 59 \cdot 3 \left(\frac{\text{lb-f}}{\text{ft}^3} \right) (51 \cdot 7 \text{ ft})$$

$$\Delta p = 3065 \cdot 8 \text{ psf}$$

or

$$\frac{3065 \cdot 8}{144} = 21 \cdot 3 \text{ psi}$$

The theoretical horsepower is

$$\frac{(51 \cdot 7)(10)}{550} = 0 \cdot 95 \text{ hp}$$

(b)

$$v = 5(1 \cdot 932) = 9 \cdot 660$$
$$Re = 5(1145) = 5725$$

This is more than 3000 and the flow is turbulent. Adding 10 per cent to the Darcy friction factor calculated for smooth pipes from eq. (3.18)

$$f = \frac{(1 \cdot 1)(0 \cdot 079)}{Re^{0 \cdot 25}} = \frac{(1 \cdot 1)(0 \cdot 079)}{5725^{0 \cdot 25}}$$

$$f = 0 \cdot 01$$

Substituting this value and the other data in eq. (3.15)

$$h_f = 4(0 \cdot 01) \left(\frac{5280}{4/12} \right) \left(\frac{9 \cdot 66^2}{64 \cdot 4} \right)$$

$$h_f = 918 \cdot 3 \text{ ft}$$
$$\Delta p = (918 \cdot 3)(59 \cdot 3) = 54,460 \text{ psf}$$

or

$$\frac{54,460}{144} = 378 \cdot 2 \text{ psi}$$

The theoretical power required is

$$\frac{(918 \cdot 3)(50)}{550} = 83 \cdot 5 \text{ hp}$$

Alternatively, since $Q = 5(0 \cdot 1686) = 0 \cdot 843$ ft³/sec,

$$\frac{(\varDelta p)\,(Q)\,(\text{lb-f/ft}^2)\,(\text{ft}^3/\text{sec})}{550(\text{ft})\,(\text{lb-f/sec})} = \frac{(54{,}460)\,(0 \cdot 843)}{550} = 83 \cdot 5 \text{ hp}$$

EXAMPLE 3.3

Water is discharged from a large reservoir through a straight pipe of 3 in. diameter and 1200 ft long at a rate of 12 cfm. The discharge end is open to the atmosphere. If the open end is 40 ft below the surface level in the reservoir, what is the Darcy friction factor? Losses other than pipe friction may be ignored.

Solution

Applying the flow equation between the levels *1* and *2*, as shown in Fig 3.5,

$$Z_1 + \frac{p_1}{\gamma} + \frac{v_1^2}{2g} = Z_2 + \frac{p_2}{\gamma} + \frac{v_2^2}{2g} + H_f$$

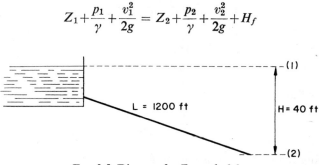

FIG. 3.5. Diagram for Example 3.3.

Both ends of the system are open to the atmosphere, then

$$p_1 = p_2$$

Ignoring losses other than pipe friction

$$H_f = h_f = 4f \left(\frac{L}{D}\right) \left(\frac{v^2}{2g}\right)$$

where v is the velocity in the pipe. It follows that $v_2 = v$, and since

$Z_1 - Z_2 = H$, and $v_1 = 0$, the flow equation assumes the form

$$H = \frac{v^2}{2g} + 4f\left(\frac{L}{D}\right)\left(\frac{v^2}{2g}\right)$$

$$H = \frac{v^2}{2g}\left(1 + 4f\frac{L}{D}\right)$$

$$Q = \frac{12}{60} = 0 \cdot 2 \text{ cfs}$$

$$A = D^2\pi/4 = (3/12)^2\pi/4 = \pi/64$$

$$v = \frac{Q}{A} = \frac{0 \cdot 2}{\pi/64} = \frac{12 \cdot 8}{\pi} \text{ fps}$$

Substituting

$$40 = \frac{(12 \cdot 8)^2}{(64 \cdot 4)\,(\pi^2)}\left(1 + 4f\frac{1200}{3/12}\right)$$

from which

$$f = 0 \cdot 008$$

3.3. The Poiseuille Equation

From eq. (3.9)

$$(\Delta p)g_c = \frac{4LR'}{D} \tag{3.19}$$

In laminar flow, from the equations (3.7) and (3.8)

$$C_D = \frac{8}{Re} = \frac{8\mu}{Dv\varrho}$$

Substituting for C_D from this equation in eq. (3.6)

$$R' = C_D\varrho v^2 = \frac{8\mu\varrho v^2}{Dv\varrho}$$

Simplifying

$$R' = \frac{8\mu v}{D} \tag{3.20}$$

From this equation and eq. (3.19)

$$(\Delta p)g_c = \left(\frac{4L}{D}\right)\left(\frac{8\mu v}{D}\right)$$

$$(\Delta p)g_c = \frac{32Lv\mu}{D^2} \tag{3.21}$$

The last equation is commonly known as the Poiseuille equation. It will be noted, however, that the original equation presented by Poiseuille (1799–1869), and another one related to the same subject, by Hagen (1797–1884), both deduced empirically, bore little resemblance to eq. (3.21). The analytical derivation of this equation was not accomplished until 1858–60, by Neumann and Hagenbach independently. The latter appears to have named eq. (3.21) after Poiseuille rather than after Hagen, and this designation has won general recognition.

Equation (3.21) can also be derived from the Darcy equation as follows. In laminar flow, from eq. (3.17)

$$f = \frac{16}{Re} = \frac{16\mu}{Dv\varrho}$$

Substituting from this equation for f in eq. (3.15)

$$h_f = 4\left(\frac{16\mu}{Dv\varrho}\right)\left(\frac{L}{D}\right)\left(\frac{v^2}{2g}\right)$$

Simplifying and taking

$$h_f = \frac{\Delta p}{\gamma} \text{ (from eq. (3.14))}$$

$$\frac{\Delta p}{\gamma} = \frac{32Lv\mu}{gD^2\varrho}$$

Substituting in this equation from eq. (1.6) for

$$\gamma = \varrho(g/g_c)$$

$$\frac{\Delta p}{\varrho(g/g_c)} = \frac{32Lv\mu}{gD^2\varrho}$$

from which

$$(\Delta p)g_c = \frac{32Lv\mu}{D^2} \tag{3.21}$$

EXAMPLE 3.4

Water (density $= 62\cdot4$ lb/ft^3, viscosity $= 0\cdot000672$ lb/(ft) (sec) flows in a tube of $0\cdot25$ in. bore at a velocity of 1 fps. What is the drop in pressure in a length of 10 ft?

Solution

$$Re = \frac{Dv\varrho}{\mu} = \frac{(0 \cdot 25)/12)\,(1 \cdot 0)\,(62 \cdot 4)}{0 \cdot 000672}$$

$$Re = 1935$$

This is less than the critical Reynolds number 2000, and eq. (3.21) applies.

$$(\Delta p)g_c = \frac{32Lv\mu}{D^2} = \frac{(32)\,(10)\,(1 \cdot 0)\,(0 \cdot 000672)}{(0 \cdot 25/12)^2}$$

$$(\Delta p)g_c = 495 \cdot 5 \text{ poundals/ft}^2$$

$$\Delta p = \frac{495 \cdot 5}{32 \cdot 2} = 15 \cdot 38 \text{ psf}$$

Alternatively, from eq. (3.16)

$$f = \frac{16}{Re} = \frac{16}{1935}$$

Substituting this value for f in eq. (3.15)

$$\frac{\Delta p}{62 \cdot 4} = 4 \left(\frac{16}{1935} \right) \left(\frac{10}{0 \cdot 25/12} \right) \left(\frac{1 \cdot 0^2}{64 \cdot 4} \right)$$

from which the pressure drop is obtained directly in engineering units

$$\Delta p = 15 \cdot 38 \text{ psf}$$

This is equivalent (from eq. (3.14)) to

$$\frac{(15 \cdot 38)\,(12)}{62 \cdot 4} = 2 \cdot 96 \text{ i.w.g.}$$

3.4. Pipe Roughness. Moody Friction Diagram

Darcy's great contribution to hydraulics lies in his conclusive demonstration of the fact that resistance to flow depends not only on the Reynolds number but also on the type and condition of the boundary material. This important fact had been overlooked in earlier studies on pipe friction, and even the so-called Darcy friction diagram only vaguely relates to this subject.

The inner surface of commercial pipes is covered with innumerable geometrical irregularities, the size, configuration and distribution of which may vary within wide limits. If the mean projected heights of such irregularities are appreciable, pipes are called rough, otherwise they are called smooth. But the roughness promotes and sustains form drag, from which one may conclude that rough pipes should produce greater frictional losses than smooth pipes of the same size. This conclusion has been amply evidenced in experimental works, but within the turbulent range only. In laminar flow the effect of roughness is negligible, as might be expected from the fact that the slow-moving laminar boundary layer within this region is rather thick. In fact, it is thick enough to cover all the possible irregularities so effectively as to make the surface exhibit the properties of a perfectly smooth pipe. This leads to the conclusion that roughness should be included as an important parameter in all correlations on pipe friction in turbulent flow.

The main problem with rough surfaces is to establish a reliable *index of roughness*. A number of investigators have attempted to solve this problem by producing some artificial roughness on the inner surface of smooth pipes. Prandtl, Schlichting, and Nikuradse studied this problem by making smooth surfaces artificially roughened using a coating of uniformly sized grains of sand. In this method the degree of roughness was known accurately, and the friction factors found experimentally were related to this roughness.

Nikuradse (1894–   ) undertook the determination of friction factors in smooth and grained pipes along with his studies of the velocity profiles in pipes. From the results obtained, Nikuradse presented the following equations for the Reynolds numbers above 5000.

For smooth pipes

$$\frac{1}{\sqrt{f'}} = -0.80 + 2.0 \log Re\sqrt{f'} \tag{3.22}$$

For rough pipes

$$\frac{1}{\sqrt{f'}} = 1.14 + 2.0 \log \frac{D}{e} \tag{3.23}$$

where e is the pipe roughness, and D its diameter. The new friction factor is defined in terms of the Darcy friction factor by

$$f' = 4f$$

The above equations confirm two interesting suppositions, namely that in smooth pipes the effect of roughness is negligible while the effect of the Reynolds number is insignificant in rough pipes at $Re > 5000$.

Stanton (1865–1931) used eq. (3.23) in presenting a friction factor diagram bearing his name. The diagram, which is not reproduced here, is a plot of the new friction factor f' vs. Re, with e/D as a parameter. The roughness–pipe diameter ratio is called the *relative roughness*.

The excellent results of Nikuradse's work suffer, however, from an inconvenient limitation, namely that—being based on artifially roughened surfaces—they are not directly reproducible in engineering problems connected with pipe friction. This limitation has been partly overcome by further investigations on uncoated commercial pipes.

From a series of tests, Colebrook evaluated the friction factors for a number of commercially important pipes, using a modified form of the Darcy equation

$$h_f = f' \frac{L}{D} \frac{v^2}{2g} \tag{3.24}$$

hence obtaining the corresponding values of their equivalent roughness from eq. (3.23). Finally, by plotting the test results, *Colebrook* obtained a single line described by the equation

$$\frac{1}{\sqrt{f'}} - 2 \log \frac{D}{e} = 1 \cdot 14 - 2 \log \left[1 + \frac{9 \cdot 28}{Re \sqrt{f'}(e/D)} \right] \tag{3.25}$$

Moody (1880–1953) extended the work of Colebrook by including information on more commercial pipes, and he also presented eq. (3.25) in the form of an earlier Stanton diagram. The diagram, reproduced in Fig. 3.6, bears his name.

In order that the two friction factors appearing in the Darcy equation can be identified, f' will be called in this text the Moody friction factor.

Following the steps leading to the eq. (3.16), eq. (3.24) may also be presented in the form

$$h_f = \frac{f' L Q^2}{40 D^5} \tag{3.26}$$

EXAMPLE 3.5

$120 \cdot 8$ hp are expended in pumping water through a horizontal pipe of 12 in. diameter, and 1000 ft long, at a rate of 600 cfm. Assuming a 70 per cent overall efficiency of the pump, what is the roughness of the pipe?

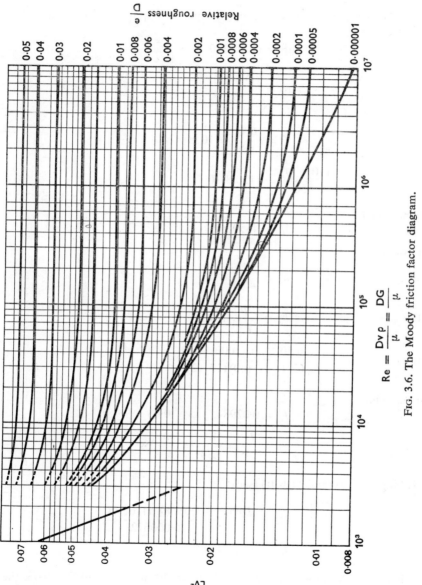

FIG. 3.6. The Moody friction factor diagram.

Solution

At 70 per cent pump efficiency, the power required to overcome pipe friction is

$$hp = (120 \cdot 8)(0 \cdot 7)$$

But also

$$hp = \frac{(Q)(\Delta p)}{550}$$

And, since

$$Q = \frac{600}{60} = 10 \text{ cfs}$$

$$\frac{(10)(\Delta p)}{550} = (120 \cdot 8)(0 \cdot 7)$$

from which

$$\Delta p = 4650 \text{ psf}$$

Using eq. (3.26)

$$h_f = \frac{f' L Q^2}{40 D^5}$$

But

$$h_f = \frac{\Delta p}{\gamma}, \quad \text{and} \quad \gamma = 62 \cdot 4 \frac{\text{lb-f}}{\text{ft}^3}$$

then

$$\frac{4650}{62 \cdot 4} = \frac{f'(1000)(10^2)}{(40)(12/12)}$$

from which

$$f' = 0 \cdot 02984, \quad \text{say} \quad 0 \cdot 03$$

$$v = \frac{Q}{A} = \frac{10}{\pi/4} = \frac{40}{\pi} \text{ fps}$$

Taking $\mu = 0 \cdot 000672$ (lb/ft) (sec)

$$Re = \frac{Dv\varrho}{\mu} = \frac{(1 \cdot 0)(40/\pi)(62 \cdot 4)}{0 \cdot 000672}$$

$$Re = 1 \cdot 183 \times 10^6$$

From the Moody diagram for this Reynolds number and $f' = 0 \cdot 03$, the relative roughness of the pipe is

$$e/D = 0 \cdot 005$$

and, since
$$D = 12 \text{ in.}$$
$$e = 0\cdot060 \text{ in.}$$

This is rather a high value for a commercial pipe, and only justified on account of very long service under extremely severe conditions.

3.5. Flow through Gradually Changing Sections

The loss of head due to friction in circular conduits of sections varying linearly with length, such as met with in Venturi tubes, for example, can be determined from any equation established for uniform pipes. These equations must, however, be handled differentially with the diameter expressed as a function of length.

Since in all practical situations the Reynolds numbers are likely to be high, even at the wider ends of the conduits, the friction factors may be assumed to be independent of this number and taken as a constant. The method of tackling problems of this kind is shown in the following numerical example.

EXAMPLE 3.6

A Venturi tube has the converging section changing in diameter from 6 to 2 in. in a length of 5 ft. What is the loss of head in this length when water flows at a rate of 1·6 cfs? The Darcy friction factor may be assumed to have a constant value of 0·006.

Solution

Consider an element of the pipe of length dx, at a distance x from its wider end, and let D be its diameter at this distance. Then, with reference to Fig. 3.7,
$$\frac{6}{2} = \frac{5+y}{y}$$
from which
$$y = 2\cdot5 \text{ ft}$$
where y is the length obtained by the sides of the pipe produced to meet on its axis.

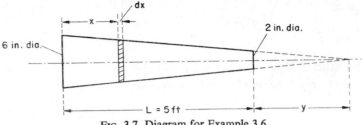

FIG. 3.7. Diagram for Example 3.6.

Also

$$\frac{D}{2/12} = \frac{(5-x)+y}{y} = \frac{(5-x)+2\cdot5}{2\cdot5}$$

from which

$$D = \frac{1}{12}\left(6 - \frac{4x}{5}\right)$$

Substituting for D from this equation, and dx for L in eq. (3.16), presented in a differential form

$$d(h_f) = \frac{fQ^2(dx)}{10(1/12)^5\,(6-4x/5)^5}$$

Integrating this equation between the limits $x = L = 5$, and $x = 0$

$$h_f = \frac{fQ^2}{10(1/12)^5} \int\limits_{x=0}^{x=5} (6-4x/5)^{-5}\,(dx)$$

$$\frac{fQ^2}{10(1/12)^5} = \frac{(0\cdot006)\,(1\cdot6)^2\,(12)^5}{10} = 382\cdot3$$

$$\int\limits_0^5 (6-4x/5)^{-5}\,(dx) = \left[\frac{6-4x/5}{-4/5(1-5)}\right]_{x=0}^{x=5} = 0\cdot0193$$

$$h_f = (382\cdot3)\,(0\cdot0193)$$
$$h_f = 7\cdot386, \quad \text{say } 7\cdot4 \text{ ft}$$

3.6. Flow through Suddenly Changing Sections

Suppose that a pipe changes its cross-sectional area suddenly, as shown in Fig. 3.8, and consider at first a sudden enlargement. The loss of energy experienced in this case is due to the eddies in the portion of the fluid trapped at the corner of the larger section of the pipe. These eddies draw appreciable energy from the stream with the result that there is a loss of head. Let this loss be h_e, then with reference to Fig. 3.8(a)

$$h_e = \frac{(v_1 - v_2)^2}{2g}$$ (3.27)

where v_1 and v_2 are the velocities in the smaller and larger sections of the pipe. Equation (3.27) has been found in fair agreement with experimental evidence, but its formal proof, based on the principle of momentum, requires a rather crude approximation, and will not be given here.

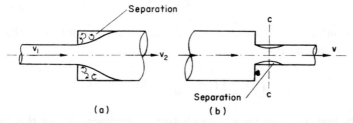

FIG. 3.8. Losses in sudden enlargement (a) and sudden contraction (b) of a section.

Now consider the flow through a pipe which suddenly reduces its cross-section, and let h_c be the loss of head due to the sudden contraction of the stream. As shown in Fig. 3.8(b), it contracts in a similar way to the stream passing through an orifice, with the effect that vena contracta forms just beyond the change in section. Experimental evidence is that the loss arising from the contraction itself is negligibly small when compared with the loss due to the enlargement of the stream following the vena contracta. On this basis, the loss is conveniently expressed by the equation

$$h_c = \frac{(v_c - v)^2}{2g}$$ (3.28)

in which v is the average velocity in the smaller pipe, and v_c is that at the vena contracta.

Let a and a_c be the respective flow areas in the smaller pipe and at the vena contracta, then from the law of continuity

$$v_c = \frac{a}{a_c}(v)$$

From the definition of the coefficient of contraction (see Chapter 1)

$$C_c = \frac{a_c}{a}$$

then, substituting in eq. (3.28), for

$$v_c = \frac{v}{C_c}$$

$$h_c = \frac{(v/c_c - v)^2}{2g}$$

from which

$$h_c = \left(\frac{1}{C_c} - 1\right)^2 \frac{v^2}{2g} \qquad (3.29)$$

For a small circular orifice, the coefficient of contraction is of the order of 0·6. Assuming this value

$$\left(\frac{1}{C_c} - 1\right)^2 = \left(\frac{1}{0·6} - 1\right)^2 = 0·44$$

Slightly higher values have been obtained experimentally for this term in pipe contraction, and 0·5 has been generally accepted as an approximation, so that

$$h_c = 0·5 \frac{v^2}{2g} \qquad (3.30)$$

Example 3.7

A process column receives a solution (viscosity = 1·5 cp, sp.gr. = 1·2) from an overhead open tank through a copper tube of 1 in. bore and 60 ft long. If the operating pressure in the column is 8 psig, what must be the height of the surface level in the tank, above the feed point at the column, in order that the flow will not fall below 90 lb/min?

Solution

With reference to Fig 3.9, let H be the required height. The flow equation, written between the surface level in the tank, (1) and the discharge point (2), assumes the form

$$\frac{p_1}{\gamma} + Z_1 + \frac{v_1^2}{2g} = \frac{p_2}{\gamma} + Z_2 + \frac{v_2^2}{2g} + H_f \qquad (1)$$

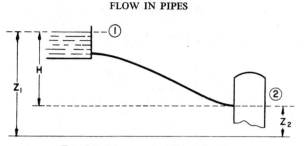

FIG. 3.9. Diagram for Example 3.7.

(In writing this equation, the correcting factor (α) has been ignored in accordance with general practice in industrial problems.)

The energy loss, in eq. (1), is composed of the loss due to the sudden contraction (h_c), at the entrance from the tank to the copper tube, the loss due to friction (h_f) in the tube, and the loss due to the sudden enlargement (h_e) at the discharge point, thus

$$H_f = h_c + h_f + h_e \qquad (2)$$

Let v be the velocity in the tube, then

$$h_c = 0.5 \frac{v^2}{2g} \qquad (3.30)$$

$$h_f = 4f \frac{L}{D} \frac{v^2}{2g} \qquad (3.15)$$

and since $v_1 = v$, and $v_2 = 0$, then from eq. (3.27)

$$h_e = \frac{v^2}{2g}$$

The density of the solution is $(1.2)(62.4)$ lb/ft^3, then for the flow of 90 lb/min

$$Q = \frac{90}{(1.2)(62.4)(60)} = 0.020 \text{ ft}^3/\text{sec}$$

The flow area corresponding to 1 in. diameter

$$A = (1/12)^2 \pi/4 = 0.00553 \text{ ft}^2$$

$$v = \frac{Q}{A} = \frac{0.020}{0.00553}$$

$$v = 3.753 \text{ fps}$$

$$Re = \frac{Dv\varrho}{\mu} = \frac{(1/12)(3.753)(1.2)(62.4)}{(1.5)(0.000672)}$$

$$Re = 23,230$$

For this Reynolds number, from Fig. 3.4, curve B, the Darcy friction factor

$$f = 0·006$$

$$h_f = 4(0·006) \cdot \frac{60}{(1/12)} \frac{v^2}{2g} = 17·28 \frac{v^2}{2g}$$

Substituting in eq. (2)

$$H_f = 0·5 \frac{v^2}{2g} + 17·28 \frac{v^2}{2g} + \frac{v^2}{2g}$$

$$H_f = 18·78 \frac{v^2}{2g}$$

From eq. (1)

$$Z_1 - Z_2 = \frac{p_2 - p_1}{\gamma} + \frac{v^2}{2g} + H_f$$

$$\frac{p_2 - p_1}{\gamma} = \frac{(8)(144)}{(1·2)(62·4)} = 15·38 \text{ ft}$$

But $Z_1 - Z_2 = H$, then

$$H = 15·38 + \frac{v^2}{2g} + 18·78 \frac{v^2}{2g}$$

$$H = 15·38 + 19·78 \frac{v^2}{2g}$$

$$19·78 \frac{v^2}{2g} = 19·78 \left(\frac{3·753^2}{64·4} \right) = 4·32 \text{ ft}$$

$$H = 15·38 + 4·32$$

$$H = 19·7 \text{ ft}$$

EXAMPLE 3.8

Two water tanks are connected by a pipeline consisting of three straight pieces, namely: 20 ft long and 2 in. diameter, 20 ft long and 4 in. diameter, and 10 ft long and 1 in. diameter, as shown in Fig. 3.9a. If the water surface in one tank is 10 ft above the surface in the other tank what will be the flow, expressed in cfm? The changes in sections are sudden, and the Darcy friction factor, $f = 0·01$, is the same for all pipes.

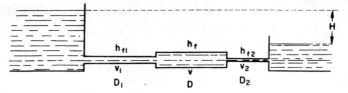

FIG. 3.9a. Diagram for Example 3.8.

Solution

With reference to Fig. 3.9a, the losses due to friction in straight pipes are

$$\Sigma h_f = h_{f1} + h_f + h_{f2} = \frac{4f}{2g}\left[\frac{L_1 v_1^2}{D_1} + \frac{L v^2}{D} + \frac{L_2 v_2^2}{D_2}\right]$$

Expressing the velocities in terms of v

$$v_1 = \left(\frac{D}{D_1}\right)^2 v = \left(\frac{4}{2}\right)^2 v = 4v$$

$$v_2 = \left(\frac{D}{D_2}\right)^2 v = \left(\frac{4}{1}\right)^2 v = 16v$$

Also putting $f = 0.01$

$$\Sigma h_f = \frac{0.04}{2g}\left[\frac{(20)(16)v^2}{2/12} + \frac{20v^2}{4/12} + \frac{(10)(256)v^2}{1/12}\right]$$

$$\Sigma h_f = 1308\frac{v^2}{2g}$$

The loss in sudden contraction between the tank and 2 in. diameter pipe (eq. (3.40))

$$h_{c1} = 0.5\frac{v_1^2}{2g} = 0.5\frac{16v^2}{2g} = \frac{8v^2}{2g}$$

The loss in sudden enlargement from 2 in. to 4 in. diameter (eq. (3.27))

$$h_{e1} = \frac{(v_1-v)^2}{2g} = \frac{(4v-v)^2}{2g} = \frac{9v^2}{2g}$$

The loss in sudden contraction from 4 in. to 1 in. diameter (eq. (3.30))

$$h_{c2} = 0.5\frac{v_2^2}{2g} = 0.5\frac{256v^2}{2g} = \frac{128v^2}{2g}$$

The loss in sudden enlargement between 1 in. diameter pipe and the tank, where the velocity is zero (eq. (3.27)),

$$h_{e2} = \frac{v_2^2}{2g} = \frac{256v^2}{2g}$$

The total losses

$$H_f = \Sigma h_f + h_{c1} + h_{e1} + h_{c2} + h_{e2} = (1308 + 8 + 9 + 128 + 256)\frac{v^2}{2g}$$

$$H_f = 1709\frac{v^2}{2g}$$

But $H_f = H = 10$, and putting $2g = 64 \cdot 4$

$$v^2 = \frac{(10)(64 \cdot 4)}{1709}$$

$$v = 0 \cdot 614 \text{ fps}$$

The flow is then

$$(60)(0 \cdot 614)\left(\frac{4}{12}\right)^2\frac{\pi}{4} = 3 \cdot 212 \text{ cfm}$$

3.7. Loss of Head in Pipe Fittings

The loss of head in flow through pipe fittings arises from the sudden change in the flow pattern which results in an excessive form drag. This kind of loss is difficult to express in a mathematical form and the common approach is to regard the fitting as an equivalent length of pipe. This

TABLE 3.1. EQUIVALENT LENGTH OF COMMON FITTINGS

Description	Equivalent number of pipe diameters (N)
90° standard elbow	30
90° long radius elbow	20
45° standard elbow	16
180° close return bend	50
Standard tee	
(a) with flow through run	20
(b) with flow through branch	60
Conventional globe valve, fully open	340
Conventional angle valve, fully open	145
Conventional gate valve, fully open	13
Butterfly valve, fully open	20
Straight-through cock	18

equivalent is then added to the actual length of the pipe for the calculation of the loss of head. Table 3.1 shows this equivalent for the most common fittings and valves, which may be regarded as special kinds of fittings.

EXAMPLE 3.8a

A solution (viscosity = 2·4 cp, sp.gr. = 1·2) is to be pumped from a storage tank to an overhead tank, both open to the atmosphere. The pipe line, 2·067 in. diameter, is 160 ft long, and the fittings include three 90° standard elbows, one tee-piece, and one conventional globe valve. The vertical distance between the surface levels in the tanks is 80 ft. If the solution is to flow through the pipe at 6 fps, what horse-power will be consumed by the pump at 50 per cent efficiency? The roughness of the pipe may be assumed to be 0·0018 in.

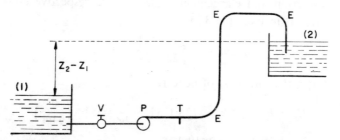

FIG. 3.10. Diagram for Example 3.8a (V = valve, P = pump, T = tee piece, E = elbow).

Solution

With reference to the Table 3.1, the equivalent length of the fittings is obtained as follows.

$$
\begin{array}{lr}
 & N \\
1 \text{ tee-piece} & = 20 \\
1 \text{ globe valve} & = 340 \\
3\times90° \text{ elbows} = 3\times30 = & 90 \\
\hline
\text{Total} & = 450
\end{array}
$$

$$ND = (450)\left(\frac{2\cdot067}{12}\right) = 77\cdot5 \text{ ft}$$

$$L = 160+77\cdot5 = 237\cdot5 \text{ ft}$$

$$Re = \frac{Dv\varrho}{\mu} = \frac{(2\cdot067)/12)(6)(1\cdot2)(62\cdot4)}{(2\cdot4)(0\cdot000672)}$$

$$Re = 48,000$$

For this Reynolds number and the relative roughness

$$e/D = 0.0018/2.067 \cong 0.0009$$

the Moody friction factor, from Fig. 3.6, is

$$f' = 0.024$$

Using eq. (3.24)

$$h_f = f' \frac{L}{D} \frac{v^2}{2g} = (0.024) \left(\frac{237.5}{2.067/12} \right) \left(\frac{v^2}{2g} \right)$$

$$h_f = 33.1 \frac{v^2}{2g}$$

where v is the velocity in the pipe.

The contraction and elargement losses, at the respective ends of the pipe line, are (as in Example 3.7)

$$h_c + h_e = 0.5 \frac{v^2}{2g} + \frac{v^2}{2g} = 1.5 \frac{v^2}{2g}$$

For $v = 6$ fps, the total loss of head is

$$H_f = h_f + h_c + h_e = (33.1 + 1.5) \frac{v^2}{2g} = 34.6 \left(\frac{6^2}{64.4} \right)$$

$$H_f = 19.34, \quad \text{say } 19.4 \text{ ft}$$

Writing the flow equation between the levels (1) and (2), in Fig. 3.10

$$Z_1 + \frac{v_1^2}{2g} + \frac{p_1}{\gamma} + W = Z_2 + \frac{v_2^2}{2g} + \frac{p_2}{\gamma} + H_f$$

Since $v_1 = v_2 = 0$, and $p_1 = p_2$, this equation reduces to

$$W = (Z_2 - Z_1) + H_f$$

where W is the net work to be done by the pump on 1 lb of the solution.

Since

$$Z_2 - Z_1 = 80 \text{ ft}$$

$$W = 80 + 19.4 = 99.4 \text{ ft}$$

or

$$99.4 \text{(ft) (lb-f)/lb-wt}$$

The cross-sectional area of the pipe is $(2.067/12)^2 \pi/4 = 0.0233$ ft², and the rate of flow is

$$(6) (0.0233) (1.2) (62.4) = 10.47 \text{ lb/sec}$$

At 50 per cent efficiency of the pump, the power consumption is

$$\frac{(99 \cdot 4)\,(10 \cdot 47)}{(550)\,(0 \cdot 5)} = 3 \cdot 78 \text{ hp}$$

3.8. Siphoning

Consider a simple example of siphoning when a liquid is drawn by means of an inverted U-tube from a higher to a lower level, as shown in Fig. 3.11.

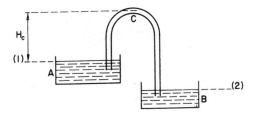

FIG. 3.11. Simple siphoning.

Let p be the atmospheric pressure, p_c the pressure at the highest point of the U-tube, and H_c the height of this point above the surface level in the reservoir A, in Fig. 3.11, then ignoring the correcting factor (α) of eq. (1.34), the following flow equation can be written.

$$\frac{p}{\gamma} = H_c + \frac{p_c}{\gamma} + \frac{v^2}{2g} + H_f \tag{3.31}$$

In this equation, v is the velocity in the tube, γ is the density of the liquid, and H_f is the head lost in friction.

If the last two terms of eq. (3.31) are very small, they may be ignored and the equation reduced to

$$\frac{p}{\gamma} = H_c + \frac{p_c}{\gamma} \tag{3.32}$$

This equation indicates that there is a limit for the height H_c at which the siphon ceases to operate. This limit is set up by the condition $p_c = 0$. This implies that (theoretically) the siphon ceases to operate when the pressure at the point c has been reduced to an absolute vacuum. At this limiting condition $p/\gamma = H_c$, and taking 14·7 psia for the atmospheric pressure, and 62·4 lb/ft³ for the density of water, the maximum theoretical height for water is

$$H_c = 34 \text{ ft approximately.}$$

In practice, this height is considerably less on account of the fact that at very low pressures water vaporises, causing discontinuity of the stream. Air, which is always present in small quantity in liquids, has an additional effect on the separation of the stream, as it tends to accumulate at the highest point of the siphon. These factors make the practical limit for the absolute pressure at c increased to about 2 psia, an equivalent of 26 in. of vacuum or 8 ft water, approximately. Next, consider siphoning by means of a long pipe, such as the one joining the two reservoirs separated by a

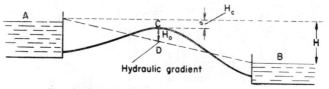

FIG. 3.12. Siphoning between large reservoirs.

hill, in Fig. 3.12. For long pipes, the velocity head term of eq. (3.31) can be ignored, and the equation written

$$H_c = \frac{p}{\gamma} - \frac{p_c}{\gamma} - H_f \qquad (3.33)$$

Assuming $p/\gamma = 34$ ft, and $p_c/\gamma = 8$ ft

$$H_c = 26 - H_f \qquad (3.34)$$

This equation indicates that for $H_f > 26$, the whole pipe will have to be below the water surface level at the higher reservoir, if separation is not to occur. From the same diagram, in Fig. 3.12, it is also evident that the vertical distance H_D of the highest point C of the pipe, measured above the hydraulic gradient (shown by the broken line), is subject to the same limitation as set out in eq. (3.34), namely that it must not exceed the assumed critical value of 26 ft, approximately.

EXAMPLE 3.9

A solution, of sp.gr. 1·2, is siphoned out of an open tank by means of a bent tube of 1 in. bore, as shown in Fig. 3.13.

The rising end of the tube is vertical and 30 ft long, while the remaining length measures 100 ft. The discharging end is open to the atmosphere and the two ends are separated by a vertical distance of 6 ft. What must be the minimum height h of the solution, measured above the dipping end of the tube, if the siphon is to be operational, and what will be the rate of flow under this limiting condition?

The Darcy friction factor is $f = 0.01$, and it may be assumed that the siphon ceases to operate when the pressure head at the highest point of the tube C falls below 6 ft of water.

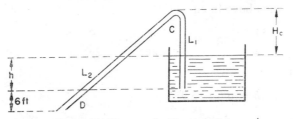

FIG. 3.13. Diagram for Example 3.9.

Solution

Making use of eq. (3.31).

$$\frac{p}{\gamma} = H_c + \frac{p_c}{\gamma} + \frac{v^2}{2g} + H_f' \qquad (1)$$

in which H_f' is the loss of head due to the contraction at the entrance to the tube (h_c), and due to friction in the rising end of the tube (h_f').

$$h_c = 0.5 \frac{v^2}{2g}$$

$$h_f' = 4f \frac{L_1}{D} \frac{v^2}{2g} = (4)(0.01)\left(\frac{30}{1/12}\right)\left(\frac{v^2}{2g}\right)$$

$$h_f' = 14.4 \frac{v^2}{2g}$$

$$H_f' = h_c + h_f' = 0.5 \frac{v^2}{2g} + 14.4 \frac{v^2}{2g}$$

$$H_f' = 14.9 \frac{v^2}{2g}$$

The critical pressure head at C is 6 ft of water, thus

$$\frac{p_c}{\gamma} = \frac{6}{1.2} = 5 \text{ ft solution}$$

Taking 34 ft water for the atmospheric pressure head

$$\frac{p}{\gamma} = \frac{34}{1\cdot 2}\text{ ft solution}$$

With reference to the diagram (Fig. 3.13)

$$H_c = 30 - h$$

Substituting the above in eq. (1)

$$\frac{34}{1\cdot 2} = (30-h)+5+\frac{v^2}{2g}+14\cdot 9\,\frac{v^2}{2g}$$

from which

$$h = 6\cdot 6+15\cdot 9\,\frac{v^2}{2g} \qquad (2)$$

Writing a flow equation between the surface level in the tank and the open end of the tube

$$\frac{p}{\gamma}+(h+6) = \frac{p}{\gamma}+H_f$$

This equation reduces to

$$h = H_f-6 \qquad (3)$$

$$H_f = h_c+h_f+h_e$$

$$h_f = 4f\frac{(L_1+L_2)}{D}\left(\frac{v^2}{2g}\right) = (4)\,(0\cdot 01)\left(\frac{130}{1/12}\right)\left(\frac{v^2}{2g}\right)$$

$$h_f = 62\cdot 4\,\frac{v^2}{2g}$$

$$h_c+h_e = 0\cdot 5\,\frac{v^2}{2g}+\frac{v^2}{2g} = 1\cdot 5\,\frac{v^2}{2g}$$

$$H_f = 1\cdot 5\,\frac{v^2}{2g}+62\cdot 4\,\frac{v^2}{2g} = 63\cdot 9\,\frac{v^2}{2g}$$

Substituting for H_f in eq. (3)

$$h = 63\cdot 9\,\frac{v^2}{2g}-6 \qquad (4)$$

Solving the eqns. (2) and (4) simultaneously

$$h = 10\cdot 8\text{ ft}$$

$$v = 4\cdot 11\text{ fps}$$

For 1-in. bore, the flow area is

$$A = (1/12)^2 \pi/4$$

$$Q = vA = \frac{(4 \cdot 11)(\pi)}{(4)(144)} \text{ cfs}$$

or

$$\frac{(4 \cdot 11)(\pi)(1 \cdot 2)(62 \cdot 4)}{(4)(144)} = 1 \cdot 68 \text{ lb/sec}$$

3.9. Parallel Flow in Pipes

Consider, in Fig. 3.14, a pipe laid in parallel along another pipe for part of its length, and joined with it at the sections (1) and (2).

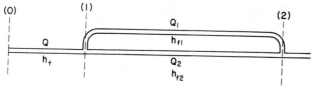

FIG. 3.14. Pipes laid in parallel.

Let h_{f1} and h_{f2} be the losses due to friction in the two parallel pipes, then, since the pressure heads (p_1/γ and p_2/γ), at the sections (1) and (2), are common to each of the two pipes

$$h_{f1} = h_{f2} = \frac{p_1 - p_2}{\gamma} \tag{3.35}$$

Ignoring losses other than pipe friction, the total loss between the sections (0) and (2) is

$$H_f = h_f + h_{f1} = h_f + h_{f2} = \frac{p_0 - p_2}{\gamma}$$

where h_f is the loss in the single pipe, and γ is the density of the flowing fluid.

Also, with reference to Fig. 3.14, $Q = Q_1 + Q_2$, and since the loss in friction is proportional to the square of velocity, it follows that laying pipes in parallel reduces the total loss in friction. This method of reducing losses is often used in engineering practice.

EXAMPLE 3.10

A small reservoir is to receive water from a lake under a differential head of 400 ft. If the conveying pipe is to be 12,000 ft long, what must be its diameter to provide a flow of 600 cfm?

If the supply is to be increased by 30 per cent, what length of additional pipe will be required to lay along the original line, both lines being of the same diameter?

The Darcy friction factor may be assumed to have the same value of 0·006 in each case.

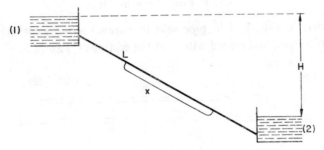

FIG. 3.15. Diagram for Example 3.10.

Solution

Writing (with reference to Fig. 3.15) a flow equation between the surface levels 1 and 2

$$\frac{p_1}{\gamma}+Z_1+\frac{v_1^2}{2g} = Z_2+\frac{v_2^2}{2g}+H_f+\frac{p_2}{\gamma}$$

But $p_1 = p_2$, $v_1 = v_2 = 0$, and $Z_1-Z_2 = H$, then

$$H = H_f$$

Let L be the length of the pipe, D its diameter, and Q the original flow, then (using eq. (3.16))

$$H = H_f = \frac{fLQ^2}{10D^5}$$

Substituting the data

$$400 = \frac{(0·006)\,(12,000)\,(600/60)^2}{10D^5}$$

from which

$$D^5 = 1·8$$

$$D = 1·125 \text{ ft}, \quad \text{or} \quad 13·5 \text{ in.}$$

Now let x be the length of the additional pipe required. The single pipe will then have the length of $12,000-x$. The new loss of head in friction taking $Q = 10+3 = 13$ ft³/sec is

$$h_f = \frac{f(12,000-x)(13)^2}{10D^5} + \frac{fx(6\cdot5)^2}{10D^5}$$

Since $h_f = H = 400$, and $D^5 = 1\cdot8$

$$400 = \frac{0\cdot006}{(10)(1\cdot8)}[(12,000-x)(13)^2+(6\cdot5)^2(x)]$$

from which

$$x = 6531 \text{ ft}$$

3.10. Branched Pipes

The solution of problems dealing with flow through branched pipes requires the writing of a number of flow equations between the branching point and liquid surface levels. The earlier assumptions, namely that the friction factor is independent of the velocity of flow and that losses other than pipe friction can be ignored, also apply here.

EXAMPLE 3.11

Two open tanks, (2) and (3) in Fig. 3.15, are to be supplied with water from a reservoir (1) through a pipe branching at (0), 50 ft below the surface level in the reservoir, the surface levels in the tanks 2 and 3 being 10 ft above and 10 ft below the junction point, respectively. The single pipe will be 200 ft long and 3·068 in. in diameter, while the branched pipes will have lengths of 100 ft and 150 ft for the higher and lower tanks, respectively. What should be the diameters of these branched pipes in order that each has the same capacity of 9 ft³/min? The Darcy friction factor is 0·008.

Solution

Ignoring the velocity head terms, the following flow equations can be written with reference to Fig. 3.16:
Between (1) and (0)

$$Z_1+\frac{p_1}{\gamma} = Z_0+\frac{p_0}{\gamma}+h_{f1}$$

But $Z_1-Z_0 = 50$, then

$$\frac{p_0-p_1}{\gamma} = 50-h_{f1} \tag{1}$$

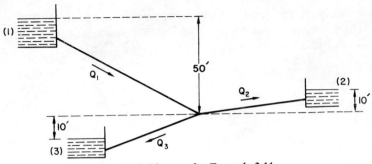

FIG. 3.16. Diagram for Example 3.11.

Between (0) and (2)

$$Z_0 + \frac{p_1}{\gamma} = Z_2 + \frac{p_2}{\gamma} + h_{f2}$$

$$Z_2 - Z_0 = 10,$$

and

$$\frac{p_0 - p_2}{\gamma} = 10 + h_{f2} \tag{2}$$

Between (0) and (3)

$$Z_0 + \frac{p_0}{\gamma} = Z_3 + \frac{p_3}{\gamma} + h_{f3}$$

$$Z_0 + Z_3 = 10$$

$$\frac{p_0 - p_3}{\gamma} = h_{f3} - 10 \tag{3}$$

$$Q_2 = Q_3 = \frac{9}{60} = 0 \cdot 15 \text{ cfs}$$

$$Q_1 = 0 \cdot 3 \text{ cfs}$$

Using eq. (3.16)

$$h_{f1} = \frac{fL_1 Q_1^2}{10 D_1^5} - \frac{(0 \cdot 008)(200)(0 \cdot 3)^2}{(10)(3 \cdot 068/12)^5}$$

$$h_{f1} = 13 \cdot 19 \text{ ft}$$

$$h_{f2} = \frac{fL_2 Q_2^2}{10 D_2^5} = \frac{(0 \cdot 008)(100)(0 \cdot 15)^2}{10 D_2^5}$$

$$h_{f2} = \frac{0 \cdot 0018}{D_2^5}$$

$$h_{f3} = \frac{fL_3 Q_3^2}{10 D_3^5} = \frac{(0 \cdot 008)(150)(0 \cdot 15)^2}{10 D_3^5}$$

$$h_{f3} = \frac{0 \cdot 0027}{D_3^5}$$

Substituting for the friction factors in the respective equations (1), (2), (3)

$$\frac{p_0 - p_1}{\gamma} = 50 - 13 \cdot 19 = 36 \cdot 81$$

$$\frac{p_0 - p_2}{\gamma} = 10 + \frac{0 \cdot 0018}{D_2^5}$$

$$\frac{p_0 - p_3}{\gamma} = \frac{0 \cdot 0027}{D_3^5} - 10$$

Since $p_1 = p_2 = p_3$ (these are all atmospheric pressures), then from 1a, and 2a

$$10 + \frac{0 \cdot 0018}{D_2^5} = 36 \cdot 81$$

from which

$$D_2^5 = \frac{0 \cdot 0018}{26 \cdot 81}$$

$$D_2 = 0 \cdot 1463 \text{ ft } (1 \cdot 756 \text{ in.})$$

From 1a and 3a

$$\frac{0 \cdot 0027}{D_3^5} - 10 = 36 \cdot 81$$

$$D_3^5 = \frac{0 \cdot 0027}{46 \cdot 81}$$

$$D_3 = 0 \cdot 1420 \text{ ft } (1 \cdot 704 \text{ in.})$$

The nearest nominal sizes would be taken for the branched pipes.

3.11. Average and Point Velocities in Pipes

Assume a steady flow in a pipe of radius r, and let $\Delta p = p_1 - p_2$ be the drop in pressure in a length L of the pipe when the average velocity of flow is v.

Now consider, in Fig. 3.17, a cylinder of the fluid of radius $y < r$. From a balance of forces

(Pressure drop) (cross-sectional area)

= (Resistance per unit area) (Surface area of the cylinder)

$$(\Delta p)(\pi y^2) g_c = (R')(2\pi y L)$$

from which

$$y(\Delta p)g_c = 2LR' \qquad (3.36)$$

where R' is the resistance per unit surface area of the cylindrical surface.

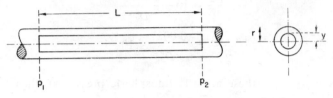

FIG. 3.17. Pressure drop and resistance to flow in pipes.

Equation (3.36) is a basic form from which a velocity distribution equation can be developed. As there is a fundamental difference in the pattern of flow between laminar and turbulent motion, these two cases will be considered separately.

(a) LAMINAR FLOW

With reference to Fig. 3.18, consider an annular element of length L formed by two coaxial fluid cylinders of radii y and $y+dy$ respectively, and let dv_x be the velocity difference across the annular element.

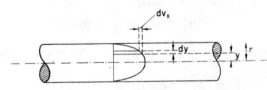

FIG. 3.18. Velocity profile in laminar flow.

In laminar flow, the resistance per unit surface area is the shearing stress as defined by eq. (1.11). It follows that

$$R' = \tau = -\mu \frac{(dv_x)}{(dy)} \qquad (3.37)$$

Substituting for R' from this equation in eq. (3.36)

$$y(\Delta p)g_c = -2L\mu \frac{(dv_x)}{(dy)}$$

Rearranging the terms

$$\frac{(\Delta p)g_c}{2L\mu}(y)(dy) = -(dv_x)$$

Integrating this equation

$$\frac{(\Delta p)g_c y^2}{4\mu L} = -v_x + C \qquad (3.38)$$

When $y = r$, $v_x = 0$, and the integration constant

$$C = \frac{(\Delta p)g_c r^2}{4\mu L}$$

Substituting for C from this equation in eq. (3.38)

$$\frac{(\Delta p)g_c y^2}{4\mu L} = -v_x + \frac{(\Delta p)g_c r^2}{4\mu L}$$

from which a point velocity v_x at a radius y is given by the equation

$$v_x = \frac{(\Delta p)g_c}{4\mu L}(r^2 - y^2) \qquad (3.39)$$

The relation between a point velocity and the average velocity in the pipe is obtained by considering an elemental flow area $dA = 2\pi y(dy)$. If the fluid flows through this area with a velocity v_x, then

$$dQ = (v_x)(2\pi y)(dy) \qquad (3.40)$$

where dQ is the elemental rate of flow.

Substituting in this equation for v_x from eq. (3.39)

$$dQ = \frac{2\pi(\Delta p)g_c}{4\mu L}(r^2 - y^2)(y)(dy) \qquad (3.41)$$

Integrating between the limits $y = r$, and $y = 0$,

$$Q = \frac{2\pi(\Delta p)g_c}{4\mu L}\left(\frac{r^4}{2} - \frac{r^4}{4}\right)$$

$$Q = \frac{\pi(\Delta p)g_c r^4}{8\mu L} \qquad (3.42)$$

But the rate of flow through a pipe of radius r is also given by $Q = \pi r^2 v$, where v is the average velocity. Equating,

$$\pi r^2 v = \frac{\pi(\Delta p)g_c r^4}{8\mu L}$$

from which

$$v = \frac{(\Delta p)g_c r^2}{8\mu L} \tag{3.43}$$

Substituting for $r = D/2$, when D is the diameter of the pipe,

$$v = \frac{(\Delta p)g_c D^2}{32\mu L} \tag{3.44}$$

The last equation will be recognised as the Poiseuille equation, usually presented in the form

$$(\Delta p)g_c = \frac{32\mu v L}{D^2} \tag{3.21}$$

Substituting in eq. (3.39), from eq. (3.43), for

$$(\Delta p)g_c = \frac{8vL\mu}{r^2}$$

$$v_x = \left[\frac{8vL\mu}{r^2}\right]\left[\frac{r^2 - y^2}{4L\mu}\right]$$

from which

$$v_x = 2v\left[1 - \left(\frac{y}{r}\right)^2\right] \tag{3.45}$$

At pipe axis, $y = 0$, and v_x has a maximum value. Let it be $v_x = v_{max}$, then from eq. (3.45)

$$v_x = v_{max} = 2v$$

or

$$\frac{v}{v_{max}} = 0.5 \tag{3.46}$$

As a matter of interest, eq. (3.46) can also be deduced from consideration of the velocity profile in laminar flow. This profile gives rise to a paraboloid the volume of which is

$$\tfrac{1}{2}(\text{base area})(\text{altitude})$$

But the altitude is v_{max}, then the average velocity

$$v = \frac{\text{volume of paraboloid}}{\text{base area}} = \frac{\tfrac{1}{2}(\pi r^2)v_{max}}{\pi r^2}$$

from which

$$\frac{v}{v_{max}} = 0.5 \tag{3.46}$$

(b) TURBULENT FLOW

Early works of Blasius, Prandtl, Karman and Nikuradse on pipe friction have resulted in a number of equations for the velocity distribution in turbulent flow. Developed largely from the theory of the boundary layer, these equations are too complicated to be directly useful in the engineering field.

Further analysis of one of the Blasius equations had led him to an assumption that the turbulent velocity profile could be approximated by an equation of the form

$$\frac{v_x}{v_{max}} = \left(1 - \frac{y}{r}\right)^n \tag{3.47}$$

where v_x is the velocity at a distance y from the centre of the pipe of a radius r, and v_{max} is the maximum velocity at the pipe axis, as shown in Fig. 3.19.

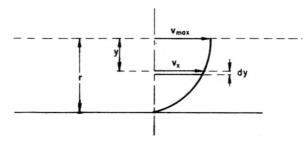

FIG. 3.19. Velocity profile in turbulent flow.

The index n of eq. (3.47) has been found experimentally to vary from $n = 1/6$ at a Reynolds number of 4000 to $n = 1/10$ at a Reynolds number of about 3×10^6. Of particular interest is its value at a Reynolds number around 10^5, when $n = 1/7$, approximately. Taking this value, eq. (3.47) becomes

$$\frac{v_x}{v_{max}} = \left(1 - \frac{y}{r}\right)^{1/7} \tag{3.48}$$

In spite of its empiricism and limitations, eq. (3.48) is very popular in engineering work where it is known as the *seventh-root* equation. On the basis of this equation, the following useful deduction can be made.

Following the line of reasoning adopted in the laminar flow, v_x can be substituted from eq. (3.48) in eq. (3.40) to give

$$dQ = (v_{\max}) \left(1 - \frac{y}{r}\right)^{1/7} (2\pi y)\,(dy)$$

$$Q = 2\pi(v_{\max}) \int_{y=0}^{y=r} \left(1 - \frac{y}{r}\right)^{1/7} (y)\,(dy) \qquad (3.49)$$

Let $z = 1 - y/r$, then $y = r(1-z)$, and $dy = -r(dz)$
When $y = r$, $z = 0$, and when $y = 0$, $z = 1$, then eq. (3.49) becomes

$$Q = -2\pi\,(v_{\max}) \int_{z=1}^{z=0} r^2\,(1-z)\,(z)^{1/7}\,(dz)$$

Changing the integration limits and rearranging the terms

$$Q = 2\pi r^2 (v_{\max}) \int_{z=0}^{z=1} (z^{1/7} - z^{8/7})\,(dz)$$

$$Q = 2\pi r^2\,(v_{\max}) \left[\frac{7}{8} z^{8/7} - \frac{7}{15} z^{15/7}\right]_{z=0}^{z=1}$$

$$Q = \pi r^2\,(v_{\max}) \left(\frac{49}{60}\right) \qquad (3.50)$$

But also

$$Q = \pi r^2 v$$

where v is the average velocity.

From the last two equations

$$\frac{v}{v_{\max}} = \frac{49}{60} = 0{\cdot}82$$

as compared with 0·5 for the same ratio in laminar flow.

EXAMPLE 3.12

Water is in laminar flow in a pipe of 1 in. radius. If at a radial distance (from the pipe axis) of 0·5 in, the point velocity is 0·1 fps, what is the Reynolds number corresponding to this flow?

Solution

Substituting the data in eq. (3.45)

$$0 \cdot 1 = 2v \left[1 - \left(\frac{0 \cdot 5}{1 \cdot 0} \right)^2 \right]$$

from which

$$v = \frac{1}{15} \, \text{fps}$$

$$Re = \frac{Dv\varrho}{\mu} = \frac{(2/12)\,(1/15)\,(62 \cdot 4)}{0 \cdot 000672}$$

$$Re = 1058.$$

3.12. Economical Pipe Diameter

For a given flow, the increase in pipe diameter reduces losses in friction but increases the cost of installation. The most economical pipe diameter may then be obtained by minimising the investment and operating charges, as illustrated in the following example.

EXAMPLE 3.13

Water is to be pumped at a rate of $1 \cdot 2$ ft³/sec. The cost of power is 3 pence/kWh and the installation charges are related to pipe diameter by

$$(40D + 2 \cdot 0) \text{ shillings/ft length}$$

where D is pipe diameter expressed in feet.

A depreciation period of 20 years may be assumed and 10 per cent would be added to the capital cost for interest and maintenance.

If the pump is to operate 3600 hr/year at 65 per cent efficiency, what will be the most economical pipe diameter? The Darcy friction factor may be assumed to have a value of $0 \cdot 01$.

Solution

The loss of head may be related to pipe diameter by eq. (3.16)

$$h_f = \frac{\Delta p}{\gamma} = \frac{fLQ^2}{10D^5} \tag{1}$$

Power absorbed in friction is

$$\frac{(\Delta p)Q}{550} = \frac{\gamma h_f Q}{550} \quad \text{(horse power)}$$

With 65 per cent efficiency, and taking 1 hp = 0·746 kW, the power consumption is

$$P = \frac{0{\cdot}746\gamma Q h_f}{(550)\,(0{\cdot}65)} \quad \text{(kW)}$$

Substituting for h_f from eq. (1)

$$P = \frac{0{\cdot}746\gamma f L Q^3}{(10)\,(550)\,(0{\cdot}65)D^5} \quad \text{(kW)}$$

Putting $Q = 1{\cdot}2$, $\gamma = 62{\cdot}4$, and $f = 0{\cdot}01$

$$P = (2{\cdot}25)\,(10^{-4})\frac{L}{D^5} \quad \text{(kW)}$$

At 3 pence/kWh, the annual cost of running the pump for 3600 hours per year, in shillings, is

$$C_1 = (2{\cdot}25)\,(3600)\left(\frac{3}{12}\right)\left(\frac{L}{D^5}\right)(10^{-4})$$

$$C_1 = (0{\cdot}20)\left(\frac{L}{D^5}\right) \tag{2}$$

The cost of installation is $(40D + 2{\cdot}0)$ shillings per foot length of pipe. Adding 10 per cent for interest and maintenance, and assuming 20 years as the depreciation period, the annual capital charges are

$$C_2 = \frac{1{\cdot}1(40D + 2)L}{20}$$

$$C_2 = L(2{\cdot}2D + 0{\cdot}11) \tag{3}$$

The total annual charges

$$C = C_1 + C_2 \tag{4}$$

$$C = L\left(\frac{0{\cdot}20}{D^5} + 2{\cdot}2D + 0{\cdot}11\right)$$

Differentiating this equation with respect to D, and equating to zero for a minimum

$$\frac{dC}{dD} = \frac{-5(0\cdot20)}{D^6} + 2\cdot2 = 0$$

$$D^6 = 0\cdot46$$

$$D = 0\cdot88 \text{ ft}$$

This is equivalent to 10·8 in., and the nearest nominal size will be taken as the most economical diameter.

Alternatively, the solution can be obtained graphically by plotting total charges against pipe diameter, as shown in Fig. 3.20. The data for the

Pipe diameter, ft	0·6	0·7	0·8	0·9	1·0	1·1
C_1/L	2·63	1·22	0·62	0·35	0·20	0·13
C_2/L	1·43	1·65	1·87	2·09	2·31	2·53
C/L	4·06	2·87	2·49	2·44	2·51	2·76

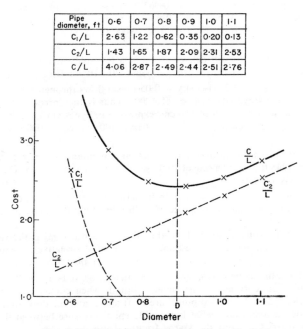

FIG. 3.20. Graphical solution for Example 3.13.

plot has been tabulated above, and the following specimen calculation refers to $D = 0\cdot6$ ft, which is the smallest pipe diameter arbitrarily chosen for the plot.

From eq. (1)

$$C_1/L = \frac{0\cdot20}{D^5} = \frac{0\cdot20}{(0\cdot6)^5}$$

$$C_1/L = 2\cdot63 \text{ shillings/ft}$$

From eq. (2)

$$C_2/L = 2 \cdot 2D + 0 \cdot 11 = (2 \cdot 2)(0 \cdot 6) + 0 \cdot 11$$
$$C_2/L = 1 \cdot 43 \text{ shillings/ft}$$

From eq. (3)

$$\frac{C}{L} = \frac{C_1}{L} + \frac{C_2}{L} = 4 \cdot 06 \text{ shillings/ft}$$

Problems

3.1. An oil (sp.gr. = 0·9, viscosity = 0·062 lb/ft×sec) flows in a pipe of 12 in. diameter at a rate of 1·5 cfs. What is the Reynolds number corresponding to this flow, and what is the loss in friction in a length of 10,000 ft, expressed in feet of oil and in psi? (*Ans.*: 1730, 20·8 ft, 8·12 psi)

3.2. What diameter pipe will deliver an oil at a Reynolds number of 2000, and at 0·1 ft^3/sec, the kinematic viscosity of the oil being $7 \cdot 0 \times 10^{-5}$ ft^2/sec? (*Ans.*: 0·91 ft)

3.3. A liquid (sp.gr. = 1·2, viscosity = 0·05 poise) flows through a horizontal tube of 0·5 cm bore at a Reynolds number of 1200. What is the corresponding velocity, and the loss of head in a length of 150 cm, expressed in dynes/cm^2, and in psi? (Take 1 lb = 454 g, 1 ft = 30·5 cm.) (*Ans.*: 100 cm/sec, 96,000 dyne/cm^2, 1·39 psi)

3.4. 85 hp are expended in overcoming friction in pumping water through a 12-in. diameter main, 1248 ft long, at a rate of 10 ft^3/sec. What is the value of the Darcy friction factor under this condition? (*Ans.*: 0·006)

3.5. Compressed air (density = 1·04 lb/ft^3, viscosity = 0·018 cp) flows through a horizontal pipe of 2 in. diameter at a rate of 50 lb/min. What Reynolds number corresponds to this flow?

If the drop in pressure along 50 ft of this pipe is 2·5 psi, what is the value of the Moody friction factor and the corresponding value of the relative roughness of the pipe? (*Ans.*: $5 \cdot 26 \times 10^5$, 0·055, 0·025)

3.6. A pipe of 2 in. diameter conveys water at an average velocity of 10 fps. A manometer containing mercury as the manometric liquid, when connected to two points of the pipe, reads 12 in. If the two points are 65 ft apart, and the downstream point is 10 ft above the upstream one, what is the differential pressure between these points, the head lost in friction, and the Darcy friction factor? (*Ans.*: 9·8 psi, 12·6 ft, 0·005)

3.7. What pipe diameter will deliver water at a rate of 37·5 ft^3/sec, if the loss of head due to friction is 215 ft of water in 8000 ft length of the pipe? The Moody friction factor is 0·0165. (*Ans.*: 1·85 ft)

3.8. Sulphuric acid (sp.gr. = 1·8) is discharged from a closed storage tank through a pipe of 1 in. diameter and 240 ft long. The discharge end of the pipe is open to the atmosphere and 26·6 ft above the acid surface in the tank.

If the air space above the acid surface in the tank is maintained at a pressure of 40 psig, what is the rate of discharge? Allow for the loss at the entrance to the pipe, and take the Darcy friction factor $f = 0 \cdot 0005$. (*Ans.*: 1·70 cfm)

3.9. Water flows through a pipe of 2 in. diameter and 25 ft length. What percentage saving in the loss of head is obtained if the central part of the pipe, 15 ft long, is replaced by a 4-in. diameter pipe of the same length? Assume sudden changes of section, and the Darcy friction factor $f = 0.01$ for both diameters. (*Ans.*: 40·4 per cent)

3.10. A flexible tube, of $\frac{1}{2}$ in. bore and 80 ft long, is used for siphoning a liquid (sp.gr. $= 0.85$) between two open tanks. The difference between the liquid surface levels in the tanks is 12 ft. The vortex of the syphon is 20 ft above the liquid surface in the higher tank and the tube leading to this point is 25 ft long. If both ends of the tube are submerged in the liquid, what is the pressure at the vortex, and the rate of syphoning? The barometric pressure is 34 ft of water and the Darcy friction factor $f = 0.01$. Allow for all losses, taking 1 atm $= 34$ ft water $= 14.7$ psi. (*Ans.*: 5·92 psi, 0·257 cfm)

3.11. Water flows through a pipe of 12 in. diameter and 7000 ft length between a lake and a reservoir, under a differential head of 40 ft. If the Darcy friction factor is $f = 0.006$, what is the rate of flow?

The highest point of the pipe is 4 ft below the water surface in the lake and the pipe leading to this point measures 5000 ft. What is the pressure head at this point, expressed in feet of water? The atmospheric pressure may be taken as an equivalent of 34 ft of water. (*Ans.*: 3·09 cfs, 8·9 ft)

3.12. Two tanks are connected by three pipes of equal length laid in parallel. Their respective diameters are D, $2D$ and $4D$. What is the discharge through each of the larger pipes if the smallest one discharges 1 cfs? Assume the friction factor to have the same value for the three pipes, and ignore losses other than pipe friction. (*Ans.*: 5·6 cfs, 32 cfs)

3.13. Water flows between two reservoirs through a pipe of 6 in. diameter and 1000 ft long, at a rate of 1·0 cfs. If the Darcy friction factor is 0·005, what is the difference between the water surface levels in the reservoirs?

Another pipe of the same length is to be laid in parallel with the existing one in order that a 50 per cent increase in the flow is attained. Assuming the same value of the friction factor, what pipe diameter is required? (*Ans.*: 16 ft, 4·55 in.)

3.14. Water flows at a constant head through two parallel pipes each of diameter D. A larger pipe is to replace the two pipes. What must be its diameter to give the same discharge? Losses other than pipe friction may be ignored, and the friction factor may be assumed the same for each pipe. (*Ans.*: 1·32 D)

3.15. A water main, 12 in. in diameter and 5000 ft long, connects two reservoirs. If the difference between the water surface levels in the reservoirs is 70 ft, and the Darcy friction factor $f = 0.005$, what is the rate of flow?

Another pipe of the same diameter and 2000 ft long is laid in parallel with the main and cross-connected, in order that the flow is increased. What percentage increase is obtained if the friction factor has the same value? (*Ans.*: 5·3 cfs, 19·3 per cent)

3.16. Water flows through a conical pipe which narrows from 12 in. to 4 in. in diameter. What is the loss of head in friction for a flow of 2·5 cfs, if the Darcy friction factor $f = 0.005$? The length of the pipe is 10 ft. (*Ans.*: 0·94 ft)

3.17. Water is to be pumped at a rate of 10 cfs.

The cost of installation is $15D^{1.5}$ shillings per foot length of the pipe, when D is the pipe diameter in ft, and 8 per cent will be added to this cost for interest and maintenance. The cost of power is 80 shillings per horse power, per year, and the overall efficiency of the pump is 70 per cent. If the depreciation period is 10 years, and the Darcy friction factor $f = 0.006$, what will be the most economical diameter of the pipe? (*Ans.*: 18·4 in.)

CHAPTER 4

OPEN CHANNELS

A CONDUIT that does not offer a solid boundary to every side of the liquid flowing in it is called an *open channel*. Rivers, canals, sewers and pipes running partly full are all examples of open channels.

The main feature of an open channel is the free surface which, as is not the case in a pipe, the liquid exhibits to the surroundings. This surface is most commonly open to the atmosphere, with the effect that the pressure on the entire open surface is atmospheric. Unless stated otherwise, the free surface will be assumed parallel with the lower boundary of the channel.

4.1. The Chèzy Equation

Early efforts, at the beginning of the eighteenth century, to grasp the principle of resistance in open channel flow, were based upon improper application of the Torricellian law of efflux. These were followed with some success by others, but the formulation of the resistance law as we know it today was left to de Chèzy (1718–88). His original investigation of the problem and its subsequent analysis have provided a background for an equation which bears his name. The equation may be derived by the application of an energy balance as follows.

Consider, in Fig. 4.1, a section of an open channel of uniform flow area laid on a slope i. Let L be a length of the channel numerically equal to the average velocity v of the liquid flowing in it, then the quantity of the liquid crossing any section of the channel in unit time is $AL\varrho$, when A is the flow area and ϱ is the density of the liquid.

The only force causing flow in open channels is the force of gravity, then, if we ignore losses other than friction, this force will be equal to the resistance experienced in the flow. This resistance may be related to the

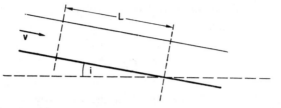

FIG. 4.1. Diagram for the derivation of the Chèzy equation.

velocity v by an equation of the form

$$R = SC_D\varrho v^2 \qquad (3.5)$$

which was derived in the preceding chapter.

Let P be the wetted perimeter of the channel, then the wetted surface area in the length considered is

$$S = PL$$

and

$$R = (PL)C_D\varrho v^2 \qquad (4.1)$$

The work done per unit time in overcoming friction is therefore

$$RL = PL^2 C_D\varrho v^2 \qquad (4.2)$$

This work equals the loss in potential energy of $AL\varrho$ mass units of the liquid when it changes its altitude by h units of length, then by an energy balance

$$AL\varrho gh = PL^2 C_D\varrho v^2$$

where g is the gravitational acceleration.

But $h = L \sin i$, and since for very small slopes normally encountered in open channel practice, we can safely take $\sin i = i$, the above equation becomes

$$AL\varrho g(Li) = PL^2 C_D\varrho v^2$$

Simplifying

$$Agi = PC_D v^2 \qquad (4.3)$$

Let

$$m = \frac{A}{P} \qquad (4.4)$$

where *m* is called the *hydraulic radius,* or *hydraulic mean depth.* Equation (4.3) may then be written

$$v^2 = \frac{gmi}{C_D} \tag{4.5}$$

This equation is normally presented in the form known as the *Chèzy equation.*

$$v = C \sqrt{(mi)} \tag{4.6}$$

It will be noted that the coefficient *C* which appears in this equation, and also known as the *Chèzy coefficient,* is given by

$$C = \sqrt{\left(\frac{g}{C_D}\right)} \tag{4.7}$$

Since the drag coefficient C_D is dimensionless, it is evident that the Chèzy coefficient has the dimension of $L^{1/2}/\theta$.

EXAMPLE 4.1

The cross-section of an open channel is shown in Fig. 4.2. If the slope of the channel bed is 1 ÷ 3600, what is the flow for a centre-line depth of

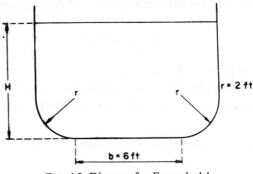

FIG. 4.2. Diagram for Example 4.1.

4 ft? What would be the depth if the flow was increased by 20 per cent? In each case, the Chèzy coefficient may be assumed to have the same value of $C = 100$ ft$^{1/2}$/sec.

Solution

With reference to Fig. 4.2, for a depth of $H = 4$ ft

$$A = 10(4-2)+(6)(2)+2\pi$$
$$A = 38 \cdot 28 \text{ ft}^2$$
$$P = 6+(2)(2)+2\pi$$
$$P = 16 \cdot 28 \text{ ft}$$

Using eq. (4.6), and taking $m = A/P$ (from eq. (4.4))

$$v = 100 \sqrt{\left(\frac{Ai}{P}\right)}$$

$$Q = vA = 100 \sqrt{\left(\frac{A^3 i}{P}\right)} = 100 \sqrt{\left(\frac{(38 \cdot 28)^3}{(16 \cdot 28)(3600)}\right)}$$

$$Q = 97 \cdot 83 \text{ cfs}$$

For $Q = (1 \cdot 2)(97 \cdot 83)$, let the unknown depth be H, then

$$A = 10(H-2)+(6)(2)+2\pi$$
$$A = 10 H - 1 \cdot 72$$
$$P = 6+2(H-2)+2\pi$$
$$P = 2H+8 \cdot 28$$

Again

$$Q = 100 \sqrt{\left(\frac{A^3 i}{P}\right)}$$

$$(1 \cdot 2)(97 \cdot 83) = 100 \sqrt{\left(\frac{(10H-1 \cdot 72)^3}{(2H+8 \cdot 28)(3600)}\right)}$$

$$\frac{(10 H - 1 \cdot 72)^3}{2H+8 \cdot 28} = 4960$$

By trial
$$H = 4 \cdot 6 \text{ ft}$$

4.2. Circular Channels

Consider, in Fig. 4.3, an open channel of circular cross-section. Let r be its radius, H be the centre depth of the liquid, and b be the breadth of its free surface. Also, let α be half of the angle subtended at the arc *egf*

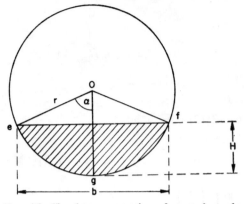

FIG. 4.3. Circular cross-section of open-channels.

which forms the wetted perimeter P, then

$$P = 2r\alpha \qquad (4.8)$$

The flow area A may be obtained by subtracting the area of the triangle oef from the area of the sector $oefg$.

$$\begin{aligned}
\text{Area of triangle} &= \tfrac{1}{2}b(r-H) \\
&= \tfrac{1}{2}(2r \sin \alpha)(r \cos \alpha) \\
&= \frac{r^2 \sin 2\alpha}{2}
\end{aligned}$$

Area of sector $= r^2\alpha$

It follows that the flow area

$$A = r^2\alpha - \frac{r^2 \sin 2\alpha}{2}$$

$$A = r^2\left(\alpha - \frac{\sin 2\alpha}{2}\right) \qquad (4.9)$$

and the hydraulic mean depth

$$m = \frac{A}{P} = \frac{r}{2}\left(1 - \frac{\sin 2\alpha}{2\alpha}\right) \qquad (4.10)$$

EXAMPLE 4.2

A circular channel of 12 in. radius is laid on a slope $1 \div 2500$. If the centre depth of water is 18 in. when the channel discharges 5 cfs, what is the value of the Chèzy coefficient?

Solution

Referring to Fig. 4.4

$$\cos \alpha = \frac{h}{r} = \frac{6}{12} = 0.5$$

$$\alpha = 60° = \frac{\pi}{3} \text{ radians}$$

Area of sector *oegf*

$$r^2\alpha = (1)^2\frac{\pi}{3} = \frac{\pi}{3} \text{ ft}^2$$

Area of triangle *oef*

$$r^2\frac{\sin 2\alpha}{2} = (1)^2\frac{\sin 120}{2} = 0.433 \text{ ft}^2$$

Area of minor segment *egf*

$$\frac{\pi}{3} - 0.433 = 0.615 \text{ ft}^2$$

Area of major segment, i.e. the *flow area*

$$A = \pi r^2 - 0.615 = 3.143 - 0.615$$
$$A = 2.528 \text{ ft}^2$$

Alternatively, using the supplementary angle $(180 - \alpha) = 120°$, and eq. (4.9)

$$A = r^2 \left(\frac{120\pi}{180} - \frac{\sin 240}{2}\right) = 2.528 \text{ ft}^2$$

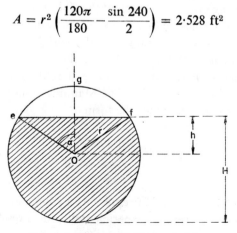

FIG. 4.4. Diagram for Example 4.2.

Also

$$P = 2r \left(\frac{120\pi}{180} \right)$$

$$P = \frac{4\pi}{3} \text{ ft}$$

$$Q = CA \sqrt{(mi)} = C \sqrt{i} \sqrt{(mA^2)}$$

$$\sqrt{(mA^2)} = \sqrt{\frac{A^3}{P}} = \sqrt{\left(\frac{(2 \cdot 528)^3}{4\pi/3} \right)} = 1 \cdot 963$$

$$5 = C(1 \cdot 963) \sqrt{1/2500}$$

$$C = 127 \cdot 3 \text{ ft}^{1/2}/\text{sec}$$

4.3. The Chèzy Coefficient

Unlike the drag factor C_D, the Chèzy coefficient (C) is dimensional. Consequently, its numerical value depends on the choice of units of length and time. Apart from this, its value depends on the shape of the channel and on the nature of its surface. To some extent, it also depends upon the velocity of flow, but the effect of this factor is too small to deserve attention in problems.

A number of attempts have been made since the end of the eighteenth century, i.e. the time when the Chèzy equation was already firmly established in engineering practice, to present the Chèzy coefficient as a function of one or more variables. While many of the resulting equations are of practical utility for restricted ranges of the variables, few if any can be accepted without some reservation. Among these, the empirical equations due to Kutter, Basin, Parkin and Manning should be mentioned as most commonly used in many countries. The Manning equation, which will be quoted here as an example, takes the form

$$C = \frac{1 \cdot 49 m^{1/6}}{N} \tag{4.11}$$

The symbol N in this equation represents a roughness factor, the value of which has been determined experimentally for a number of constructional materials, as shown in Table 4.1.

TABLE 4.1

$$\text{Typical values of } N, \frac{\text{ft}^{1/6}}{\text{ft}^{1/2}/\text{sec}}$$

Smooth brass, glass and asphalted surface	0·0005–0·009
Old wood and rusted steel	0·005 –0·015
Concrete	0·013 –0·016
Cast iron	0·014 –0·017
Riveted steel	0·017 –0·020
Brick	0·012 –0·020
Earth	0·020 –0·030
Gravel	0·022 –0·035

EXAMPLE 4.3

A concrete-lined rectangular channel, 40 ft wide, is laid on a slope of $1 \div 10,000$. What is the depth of water in the channel for a flow of 2133 cfs?

Solution

Using eq. (4.11), and taking $N = 0·015$

$$C = \frac{1·49 \, m^{1/6}}{0·015} = 100 \, m^{1/6} \tag{1}$$

$$\sqrt{i} = \sqrt{0·0001} = 0·01$$

$$Q = CA\sqrt{(mi)} = A\sqrt{m}.(0·01)(100 \, m^{1/6})$$

$$Q = Am^{2/3} \tag{2}$$

Let H be the unknown depth, then for a channel of rectangular cross-section

$$A = bH = 40H$$

$$P = b+2H = 40+2H$$

$$m = \frac{A}{P} = \frac{40H}{40+2H} = \frac{20H}{20+H}$$

Substituting for $Q = 2133$ cfs and the other data in eq. (2)

$$2133 = 40H \left[\frac{20H}{20+H} \right]^{2/3}$$

From which by trial

$$H = 13\cdot33 \text{ ft.}$$

4.4. Optimisation in Open Channels

For a given flow, an open channel can be made either broad and shallow or deep and narrow, or have any intermediate dimensions. In each case the amount of excavation work involved, hence the capital investment, is different. The most economical dimensions can be obtained by the process of optimisation as follows.

For an open channel

$$Q = CA\sqrt{(mi)} \tag{4.12}$$

Assuming that eq. (4.11) is applicable

$$Q = \frac{1\cdot49}{N} A\sqrt{(i)}m^{2/3} \tag{4.13}$$

For a given channel, N and i are constants, and we may write

$$Q = kAm^{2/3} \tag{4.14}$$

where k is a constant.

Substituting in this equation for $m = A/P$

$$Q = kA^{5/3}P^{-2/3} \tag{4.15}$$

In channels of non-circular section, the optimisation problem is confined to the determination of the best proportions of the channel, on the assumption that the flow area A is a constant. This is equivalent to the determination of the proportions for a maximum discharge through a given flow area.

4.5. Optimum Proportions of Rectangular Channels

Let (in Fig. 4.5) b be the breadth of a rectangular channel, and H be the depth of flow, then

$$A = bH, \quad \text{or} \quad b = \frac{A}{H}$$

and

$$P = b + 2H = \frac{A}{H} + 2H$$

Substituting from this equation for P in eq. (4.15)

$$Q = kA^{5/3} \left(\frac{A}{H} + 2H \right)^{-2/3}$$

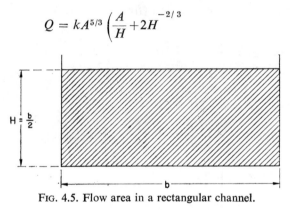

$$H = \frac{b}{2}$$

$$b$$

Fig. 4.5. Flow area in a rectangular channel.

Since A is a constant, then differentiating this equation with respect to H and equating to zero for a maximum

$$\frac{dQ}{dH} = -\frac{2}{3} kA^{5/3} \left(\frac{A}{H} + 2H \right)^{-5/3} \left(-\frac{A}{H^2} + 2 \right) = 0$$

$$-\frac{A}{H^2} + 2 = 0$$

$$2 = \frac{A}{H^2} = \frac{bH}{H^2} = \frac{b}{H}$$

$$H = \frac{b}{2} \tag{4.16}$$

It follows that the most economical proportions of a rectangular channel are when the depth of flow equals half of the breadth of the channel.

EXAMPLE 4.4

A rectangular channel is to be excavated on a slope of $1 \div 10{,}000$. What would be the most economical dimensions of the channel for a flow of 8 cfs? The Chèzy coefficient may be assumed constant at $C = 100 \text{ ft}^{1/2}/\text{sec}$.

Solution

Using eq. (4·6)

$$Q = Av = CA\sqrt{(mi)}$$

For the most economical excavation (from eq. (4.16))

$$H = \frac{b}{2}$$

$$A = bH = \frac{b^2}{2}$$

$$P = b + 2H = 2b$$

$$A\sqrt{m} = A\sqrt{\frac{A}{P}} = \sqrt{\frac{A^3}{P}} = \sqrt{\left(\frac{b^6}{(2^3)(2b)}\right)}$$

$$A\sqrt{m} = \sqrt{\frac{b^5}{16}}$$

$$Q = CA\sqrt{(mi)} = C\sqrt{i}\left(\sqrt{\frac{b^5}{16}}\right)$$

Substituting the data

$$8 = 100\sqrt{0 \cdot 0001}\left(\sqrt{\frac{b^5}{16}}\right)$$

$$\sqrt{b^5} = 32$$

$$b = 4 \text{ ft}$$

and (from eq. (4.16))

$$H = 2 \text{ ft}$$

4.6. Optimum Proportions of Trapezoidal Channels

Let (in Fig. 4.6) b be the breadth of the bottom of the trapezoidal channel, α be the angle which the sides of the channel make with the bottom, and H be the liquid depth. Let also the horizontal projection of the wetted

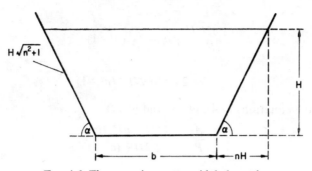

FIG. 4.6. Flow area in a trapezoidal channel.

side be nH, then

$$\tan \alpha = \frac{H}{nH} = \frac{1}{n} \tag{4.17}$$

Also, with reference to the diagram in Fig. 4.6

$$A = (b+nH)H \tag{4.18}$$
$$P = b+2H\sqrt{(n^2+1)} \tag{4.19}$$

Substituting from eq. (4.18) for

$$b = \frac{A}{H}-nH \tag{4.20}$$

in eq. (4.19)

$$P = \frac{A}{H}-nH+2H\sqrt{(n^2+1)}$$

Again, substituting from this equation for P in eq. (4.15)

$$Q = kA^{5/3}\left[\frac{A}{H}-nH+2H\sqrt{(n^2+1)}\right]^{-2/3}$$

Differentiating this equation with respect to H and equating to zero for a maximum

$$\frac{dQ}{dH} = -\frac{2}{3}kA^{5/3}\left[-\frac{A}{H^2}-n+2\sqrt{(n^2+1)}\right]\left[\frac{A}{H}-nH+2H\sqrt{(n^2+1)}\right]^{-5/3} = 0$$

$$-\frac{A}{H^2}-n+2\sqrt{(n^2+1)} = 0$$

Substituting in this equation for A from eq. (4.18), and regrouping the terms

$$\frac{(b+nH)H}{H^2}+n = 2\sqrt{(n^2+1)}$$

from which

$$b+2nH = 2H\sqrt{(n^2+1)} \qquad (4.21)$$

But from the equations (4.4), (4.18) and (4.19)

$$m = \frac{A}{P} = \frac{(b+nH)H}{b+2H\sqrt{(n^2+1)}}$$

Substituting for $2H\sqrt{(n^2+1)}$ in this equation from eq. (4.21)

$$m = \frac{(b+nH)H}{b+(b+2nH)}$$

Simplifying

$$m = \frac{H}{2} \qquad (4.22)$$

Equation (4.22) is the mathematical statement of the conditions for the most economical proportions of a trapezoidal channel.

EXAMPLE 4.5

A trapezoidal channel is to have a flow area of 120 ft². What will be the most economical dimensions of this channel if its sides make an angle of 45° with its lower boundary?

Solution

With reference to Fig. 4.6, for $\alpha = 45°$

$$\tan \alpha = \frac{1}{n} = 1$$

$$n = 1$$

Using eq. (4.19)

$$P = b+2H\sqrt{(n^2+1)} = b+2\sqrt{2}H$$

For the most economical proportions of the channel, from eq. (4.22)

$$m = \frac{A}{P} = \frac{H}{2}$$

from which

$$A = \frac{H}{2}(P) = \frac{H}{2}(b+2\sqrt{2}H)$$

But

$$A = (b+H)H \tag{4.18}$$

Equating

$$(b+H)H = \frac{H}{2}(b+2\sqrt{2}H)$$

from which

$$\frac{b}{2} = (\sqrt{(2)}-1)H = 0{\cdot}415H$$

$$b = 0{\cdot}83H$$

Substituting for b from this equation and for $A = 120$ ft² in eq. (4.18)

$$120 = (0{\cdot}83H+H)H$$
$$H^2 = 65$$
$$H = 8{\cdot}1 \text{ ft}$$
$$b = 6{\cdot}7 \text{ ft}$$

4.7. Maximum Discharge through a Circular Section

For a given radius of a circular section, the maximum discharge is re-lated to the depth of flow by eq. (4.25), which can be obtained as follows.
From eq. (4.15)

$$Q = k\sqrt[3]{(A^5/P^2)}$$

Referring to Fig. 4.3, let α be half of the angle subtended at the wetted perimeter, then for Q_{max}, A^5/P^2 is a maximum, and

$$\frac{d(A^5/P^2)}{d\alpha} = \left[P^2(5A^4)\frac{dA}{d\alpha} - A^5(2P)\frac{dP}{d\alpha} \right] \frac{1}{P^4} = 0$$

$$P^2(5A^4)\frac{dA}{d\alpha} - A^5(2P)\frac{dP}{d\alpha} = 0$$

Simplifying

$$5P\frac{dA}{d\alpha} - 2A\frac{dP}{d\alpha} = 0 \tag{4.23}$$

But

$$P = 2r\alpha \tag{4.8}$$

and

$$A = r^2\left(\alpha - \frac{\sin 2\alpha}{2}\right) \tag{4.9}$$

then

$$\frac{dP}{d\alpha} = 2r$$

and

$$\frac{dA}{d\alpha} = r^2(1 - \cos 2\alpha)$$

Substituting from these equations in eq. (4.23)

$$5(2r\alpha)\,[r^2(1 - \cos 2\alpha)] - \left[2r^2\left(\alpha - \frac{\sin 2\alpha}{2}\right)(2r)\right] = 0$$

from which

$$3\alpha - 5\alpha \cos 2\alpha + \sin 2\alpha = 0$$

By trial

$$\alpha = 154° \tag{4.24}$$

Also, since

$$H = r(1 - \cos \alpha) = r(1 - \cos 154)$$
$$H = 1\cdot9r \tag{4.25}$$

It follows that for maximum flow in a circular channel the centre depth should be 1·9 times the radius of the channel.

EXAMPLE 4.6

A circular channel is required to discharge 84 cfs of water at a slope of 1 ÷ 2500. What will be the most economic radius of the channel, if it is lined with concrete?

Solution

From Table 4.1, for concrete, $N = 0\cdot015$, approximately, and from eq. (4.11)

$$C = \frac{1\cdot49m^{1/6}}{N} = \frac{1\cdot49m^{1/6}}{0\cdot015}$$

$$C = 100m^{1/6}$$

$$Q = CA\sqrt{(mi)} = 100m^{1/6}A\sqrt{m(\sqrt{1/2500})}$$

$$Q = 2Am^{2/3} = 2A(A/P)^{2/3}$$

$$\frac{Q}{2} = \sqrt[3]{(A^5/P^2)} \tag{1}$$

For a maximum discharge

$$\alpha = 154° = \frac{154\pi}{180} \quad \text{radians}$$

From eq. (4.8)

$$P = 2r\alpha = 2\left(\frac{154\pi}{180}\right)r$$

$$P = 5\cdot374r$$

Also from eq. (4.9)

$$A = r^2\left(\alpha - \frac{\sin 2\alpha}{2}\right) = r^2\left(\frac{154\pi}{180} - \frac{\sin 308}{2}\right)$$

$$A = 3\cdot085r^2$$

$$\frac{Q}{2} = \sqrt[3]{\left(\frac{(3\cdot085)^5 r^{10}}{(5\cdot374)^2 r^2}\right)}$$

$$\left(\frac{84}{2}\right)^3 = \frac{(3\cdot085)^5 r^8}{(5\cdot374)^2}$$

$$r = \left[\frac{(42)^3 (5\cdot374)^2}{(3\cdot085)^5}\right]^{1/8}$$

$$r = 3\cdot07 \text{ ft}$$

4.8. Weirs and Notches

Weirs and notched weirs, or simply notches, are the metering devices for open channels. They are specified as sharp or broad crested, according to the width of the crest, as shown in Fig. 4.7.

A general equation for a sharp-crested weir, or notch, may be derived with reference to Fig. 4.8. The diagram shows a sharp-crested weir laid across a channel conveying a liquid at a uniform velocity v_0. As the liquid approaches the weir this velocity is still retained at the

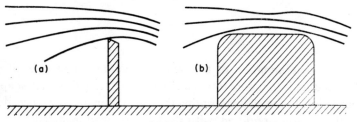

(a)   (b)

FIG. 4.7. Sharp-crested (a) and broad-crested (b) weirs.

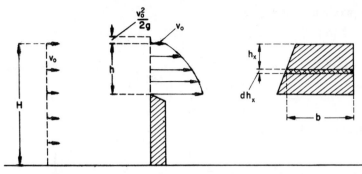

FIG. 4.8. Diagram for the derivation of an equation for a sharp-crested weir, or notch.

free surface, but it gradually increases with depth. At the weir itself, the velocity profile is shown in Fig. 4.8.

In Fig. 4.8, let h be the depth of the liquid measured at the crest, and h_x be the pressure head above a horizontal strip of the liquid of height dh_x and of a width b, then the volumetric rate of flow across the strip is

$$dQ = b(dh_x)\sqrt{(2gh_x)} \tag{4.26}$$

The total flow over the weir is therefore

$$Q = \sqrt{(2g)} \int_{h_x=h_0}^{h_x=h+h_0} bh_x^{1/2}(dh_x) \tag{4.27}$$

where the velocity head

$$h_0 = \frac{v_0^2}{2g} \tag{4.28}$$

Now, let B be the width of the channel and H be the depth of the liquid upstream from the weir; then the velocity of approach

$$v_0 = \frac{Q}{BH}$$

Since the determination of this velocity requires knowledge of Q, the usual procedure is to ignore this velocity in calculating Q, as a first approximation. This is followed by repeating the calculation once or twice, depending on the accuracy required, with the velocity head included.

4.9. Rectangular Weirs and Notches

For a rectangular weir or notch, the width (b) is a constant, and eq. (4.27) becomes

$$Q = b\sqrt{(2g)}\int_{h_x=h_0}^{h_x=h+h_0} h_x^{1/2}(dh_x) \qquad (4.29)$$

Integrating this equation, and introducing the coefficient of discharge (C_D)

$$Q = \tfrac{2}{3}C_D b\sqrt{(2g)}[(h+h_0)^{3/2} - h_0^{3/2}] \qquad (4.30)$$

The value of the coefficient of discharge is of the order of 0·6. Taking this figure

$$\tfrac{2}{3}C_D\sqrt{(2g)} = \tfrac{2}{3}(0\cdot6)\sqrt{64\cdot4} = \tfrac{10}{3} = 3\cdot33$$

and eq. (4.30) may be written

$$Q = \tfrac{10}{3}b[(h-h_o)^{3/2} - h_o^{3/2}] \qquad (4.31)$$

Otherwise, ignoring the velocity of approach

$$Q = \tfrac{10}{3}bh^{3/2} \qquad (4.32)$$

Some refinement of this equation is necessary to allow for the reduction of the actual flow area due to the contractions at the sides of the weirs, and particularly of small notches. From experimental evidence, Francis (1815–92) has shown that the side contractions reduce the effective width of a weir by a factor of (0·1 nh), in which n is the number of the contractions. Making allowance for this effect, eq. (4.32) becomes

$$Q = 3\cdot33(b-0\cdot1nh)h^{3/2} \qquad (4.33)$$

Equation (4.33) is known as the Francis formula.

EXAMPLE 4.7

A rectangular notch, 6 ft wide, is used for measuring the flow of water in a semicircular channel of 10 ft radius. If the measured depth above the crest of the notch is 4 ft, while the centre depth in the channel is 8 ft, what is the flow? The side contractions may be ignored.

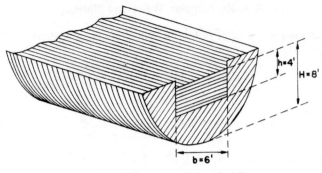

FIG. 4.9. Diagram for Example 4.7.

Solution

Using eq. (4.32) as a first approximation

$$Q = \tfrac{10}{3}bh^{3/2} = \tfrac{10}{3}(6)\,(4)^{3/2}$$

$$Q = 160 \text{ cfs}$$

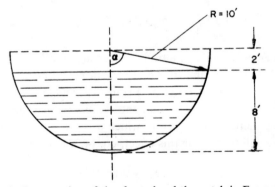

FIG. 4.10. Cross-section of the channel and the notch in Example 4.7.

With reference to Fig. 4.10

$$\cos \alpha = \tfrac{2}{10}$$

$$\alpha = 78°28', \quad \text{say} \quad 78\cdot5°$$

$$A = R^2\left(\alpha - \frac{\sin 2\alpha}{2}\right) = 10^2\left(\frac{78\cdot5\pi}{180} - \frac{\sin 157}{2}\right)$$

$$A = 117\cdot5 \text{ ft}^2$$

The velocity of approach

$$v_0 = \frac{Q}{A} = \frac{160}{117 \cdot 5} \quad \text{fps}$$

The head due to this velocity (eq. (4.28))

$$h_0 = \frac{v_0^2}{2g} = \frac{(160/117 \cdot 5)^2}{64 \cdot 4}$$

$$h_0 = 0 \cdot 03 \text{ ft, approx.}$$

Using eq. (4.31)

$$Q = \frac{10}{3} (b)[(h+h_0)^{3/2} - h_0^{3/2}]$$

$$= \frac{(10)(6)}{3} [(4+0\cdot03)^{3/2} - (0\cdot03)^{3/2}]$$

$$Q = 161 \cdot 8 \text{ cfs.}$$

4.10. V-notch

Unlike rectangular notches, notches of triangular section, commonly called V-notches, are not suitable for large discharges. The V-notch has, however, an advantage over a rectangular notch that it does not suffer from side contractions, and consequently its coefficient of discharge is independent of the rate of flow.

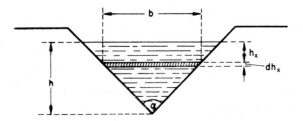

Fig. 4.11. Diagram for the derivation of a V-notch equation.

Consider, in Fig. 4.11, the cross-section of a notch of angle α. Let h be the depth of liquid measured above its apex, and h_x be the depth of a horizontal strip of height dh_x and of a width b, then from the geometry of the triangle

$$b = 2(h-h_x) \tan \frac{\alpha}{2}$$

Substituting for b from this equation in eq. (4.26)

$$dQ = 2(h-h_x) \tan \frac{\alpha}{2} (dh_x) \sqrt{(2gh_x)}$$

Integrating this equation, and introducing the coefficient of discharge (C_D)

$$Q = 2C_D \sqrt{(2g)} \tan \frac{\alpha}{2} \int_0^h (h-h_x)h_x^{1/2}(dh_x)$$

$$Q = 2C_D \sqrt{(2g)} \tan \frac{\alpha}{2} \left[\frac{2}{3} hh_x^{3/2} - \frac{2}{5} h_x^{5/2} \right]_0^h$$

$$Q = 2C_D \sqrt{(2g)} \tan \frac{\alpha}{2} \left[\frac{2}{3} h^{5/2} - \frac{2}{5} h^{5/2} \right]$$

$$Q = \frac{8}{15} C_D \sqrt{(2g)} \tan \frac{\alpha}{2} h^{5/2} \qquad (4.34)$$

For a 90° notch, commonly used in laboratory practice, $\tan \alpha/2 = 1$, and taking $C_D = 0\cdot6$, eq. (4.34) simplifies to

$$Q = 2\cdot56 \, h^{5/2} \qquad (4.35)$$

EXAMPLE 4.8

A 90° V-notch measures the flow of water in a rectangular channel, 3 ft wide. If the depth of water in the channel is 2 ft and that measured above the apex of the notch is 1 ft, what is the slope of the channel for a Chèzy coefficient of 100 ft$^{1/2}$/sec?

Solution

For the 90° notch (eq. (4.35))

$$Q = 2\cdot56 \, h^{5/2} = 2\cdot56 \, (1)^{5/2}$$

$$Q = 2\cdot56 \text{ cfs}$$

For the rectangular channel

$$A = bH = (3)(2) = 6 \text{ ft}^2$$
$$P = b+2H = 3+4 = 7 \text{ ft}$$
$$Q = CA\sqrt{\left(\frac{A}{P}i\right)} = C\sqrt{\left(\frac{A^3}{P}i\right)}$$
$$2\cdot56 = 100\sqrt{\left(\frac{6^3 i}{7}\right)}$$
$$i = \left(\frac{2\cdot56}{100}\right)^2\left(\frac{7}{6^3}\right)$$
$$i = 1 \div 39,000$$

4.11. Flow at Variable Depth

Consider, in Fig. 4.12, a free surface curving moderately upwards within a short horizontal distance dL. Let the average velocity change within this distance from v to $v+dv$ (where dv is negative in this particular case), when

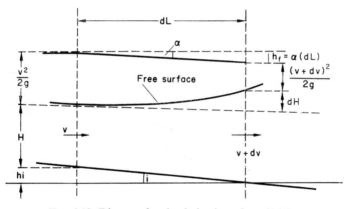

FIG. 4.12. Diagram for the derivation of eq. (4.39).

the depth changes from H to $H+dH$, then the following balance can be set up:

$$h_i+H+\frac{v^2}{2g} = (H+dH)+\frac{(v+dv)^2}{2g}+h_f \qquad (4.36)$$

For small angles, one can safely take $h_i = i(dL)$, and $h_f = \alpha(dL)$, then

$$i(dL)+H+\frac{v^2}{2g} = (H+dH)+\frac{(v+dv)^2}{2g}+\alpha(dL)$$

Simplifying and rearranging the terms

$$(i-\alpha)\,dL = dH + \frac{v(dv)}{g} + \frac{(dv)^2}{2g}$$

Dividing this equation by dL, and ignoring the second order term $(dv)^2/2g$, as negligibly small

$$i-\alpha = \frac{dH}{dL} + \frac{v(dv)}{g(dL)} \qquad (4.37)$$

The last term of this equation can be expanded as follows:

$$v = \frac{Q}{A}, \quad \text{and} \quad dv = -\frac{Q}{A^2}(dA)$$

$$\frac{dv}{dL} = \left(\frac{dv}{dH}\right)\left(\frac{dH}{dL}\right) = -\frac{Q}{A^2}\left(\frac{dA}{dH}\right)\left(\frac{dH}{dL}\right)$$

Assuming a channel of rectangular section

$$A = bH, \quad \text{and} \quad \frac{dA}{dH} = b,$$

then

$$\frac{dv}{dL} = -\frac{Qb(dH)}{A^2(dL)}$$

or

$$\frac{dv}{dL} = -\left(\frac{vb}{A}\right)\left(\frac{dH}{dL}\right)$$

Substituting for dv/dL from this equation in eq. (4.37)

$$i-\alpha = \frac{dH}{dL} + \frac{v}{g}\left[-\frac{vb(dH)}{A(dL)}\right]$$

$$i-\alpha = \frac{dH}{dL}\left(1 - \frac{v^2 b}{gA}\right)$$

For a rectangular channel

$$\frac{v^2 b}{gA} = \frac{v^2 b}{gbH} = \frac{v^2}{gH}$$

hence

$$i-\alpha = \frac{dH}{dL}\left(1 - \frac{v^2}{gH}\right) \qquad (4.38)$$

The rate of change of the slope of the free surface is then given by

$$\frac{dH}{dL} = \frac{i-\alpha}{1 - v^2/gH} \qquad (4.39)$$

The dimensionless term v^2/gH, which appears in this equation, or its square root, is called the Froude number.

Six interesting limits can be deduced from eq. (4.39).

The limit zero is approached when $\alpha \to i$, and $v^2/gH < 1$. The free surface is then parallel with the lower boundary of the channel, and the flow is said to be normal.

The second limit is set up by the condition $i > \alpha$, and $v^2/gH \to 1$, when the denominator of eq. (4.39) approaches zero, and $dH/dL \to \infty$. This limit requires that the free surface rises abruptly, the condition for which eq. (4.39) does not hold. Though physically impossible, this limit explains the tendency of the free surface to rise when the limit is approached. The phenomenon is known as the *hydraulic jump*.

The remaining four limits are stipulated by the inequalities

$$i \gtrless \alpha$$

$$\frac{v^2}{gH} \gtrless 1$$

The four possible combinations result in dH/dL to be either positive or negative, thus indicating the kind of profile of the free surface.

In variable depth flow, the loss due to friction may be conveniently related to the Chèzy coefficient by eq. (4.40), which can be derived as follows.

By analogy to flow in pipes

$$dh_f = \alpha(dL) = 4f \frac{(dL)}{D} \left(\frac{v^2}{2g} \right)$$

where f is the Darcy friction factor.

For a pipe of diameter D

$$m = \frac{A}{P} = \frac{\pi/4 D^2}{\pi D} = \frac{D}{4}$$

$$D = 4m$$

$$\alpha(dL) = 4f \frac{(dL)}{4m} \left(\frac{v^2}{2g} \right)$$

from which

$$\alpha = \frac{fv^2}{2mg}$$

From the Chézy equation (eq. (4.6))

$$v = C\sqrt{(mi)}$$

in which
$$C = \sqrt{(g/C_D)} = \sqrt{(2g/f)}$$

$$f = \frac{2g}{C^2}$$

$$\alpha = \frac{fv^2}{2mg} = \left(\frac{2g}{C^2}\right)\left(\frac{v^2}{2mg}\right)$$

$$\alpha = \frac{v^2}{mC^2} \qquad (4.40)$$

EXAMPLE 4.9

A rectangular channel, 10 ft wide, and laid on a slope of 3 in 800, carries water at a rate of 125 cfs. A weir raises the depth of the water to 5 ft, measured shortly upstream from the weir. If the Chèzy coefficient is 100 ft$^{1/2}$/sec, what is the depth 100 ft upstream from the weir?

Solution

Referring to the section shortly upstream from the weir

$$A = (10)(5) = 50 \text{ ft}^2$$

$$P = 10+2(5) = 20 \text{ ft}$$

$$m = AP = 2\cdot5 \text{ ft}$$

$$v = \frac{Q}{A} = \frac{125}{50} = 2\cdot5 \text{ fps}$$

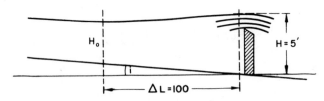

FIG. 4.13. Diagram for Example 4.9.

Using eq. (4.40)

$$\alpha = \frac{v^2}{mC^2} = \frac{(2\cdot5)^2}{(2\cdot5)\,(10{,}000)}$$

$$\alpha = 0\cdot00025$$

$$i = 3/800 = 0\cdot00375$$

$$i-\alpha = 0\cdot0035$$

$$\frac{v^2}{gH} = \frac{(2\cdot5)^2}{(32\cdot2)\,(5)} = \frac{2\cdot5}{64\cdot4}$$

$$\frac{dH}{dL} = \frac{i-\alpha}{1-v^2/gH} = \frac{0\cdot0035}{1-2\cdot5/64\cdot4} = 0\cdot00365$$

For small changes in the slope, the free surface may be assumed to curve uniformly, and

$$\frac{dH}{dL} = \frac{\Delta H}{\Delta L} = 0\cdot00365$$

For

$$\Delta L = 100 \text{ ft}$$

$$\Delta H = 0\cdot365 \text{ ft}$$

The depth, 100 ft upsteam from the weir, is then

$$H_0 = 4\cdot635 \text{ ft}$$

4.12. Specific Head

The total head over a given section of a channel is defined as the sum of the depth of the liquid, the velocity head and the potential head. Since the last quantity is proportional to the slope of the channel's bed, which is a constant, it is often convenient to ignore it in analytical considerations. The resulting equation is

$$H_s = H + \frac{v^2}{2g} \tag{4.41}$$

in which the term H_s is commonly called the *specific head* or *specific energy*. For a steady flow

$$Q = vA = \text{constant}$$

and

$$\frac{v^2}{2g} = \frac{Q^2}{2gA^2} \tag{4.42}$$

hence

$$H_s = H + \frac{Q^2}{2gA^2} \tag{4.43}$$

As the flow area is a function of depth, it is convenient at this point to refer to a specific section of the channel. For the sake of simplicity let this section be rectangular, then

$$A = bH,$$

and from eq. (4.33)

$$H_s = H + \left(\frac{Q^2}{2gb^2}\right)\left(\frac{1}{H^2}\right) \tag{4.44}$$

or

$$Q = \sqrt{\{2gb^2(H_sH^2 - H^3)\}} \tag{4.45}$$

The plots of H_s and Q from these equations against H produce the curves shown in the diagrams below.

From the diagrams, in Fig. 4.14, it may be observed that for any value of H_s or Q, two values of H are possible, except for a condition which may be described as critical. At this condition, the specific head is a minimum

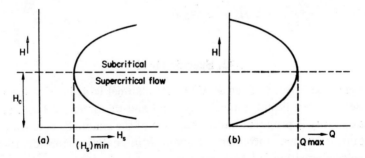

FIG. 4.14. The specific head (a), and the flow (b), as a function of depth in channel flow.

and the flow is a maximum. The depth and velocity, corresponding to this condition are said to be critical, while the velocity above and below the critical velocity is called *supercritical* or rapid, and *subcritical* or tranquil, respectively.

The following interesting conclusion can be drawn by differentiating the last two equations and equating to zero. From eq. (4.44)

$$\frac{dH_s}{dH} = 1 - \frac{2Q^2}{2gb^2H^3} = 0$$

$$H = \sqrt[3]{\left(\frac{Q^2}{gb^2}\right)} \qquad (4.46)$$

For a rectangular channel, $Q = vbH$, then

$$H = \sqrt[3]{\left(\frac{v^2H^2}{g}\right)} \qquad (4.47)$$

Let the subscript c refer to the critical condition, then

$$H_c = \sqrt[3]{\left(\frac{v_c^2H_c^2}{g}\right)}$$

from which

$$H_c = \frac{v_c^2}{g} \qquad (4.48)$$

or

$$\frac{v_c^2}{gH_c} = 1 \qquad (4.49)$$

The last equation stipulates the condition for the minimum specific head or the maximum flow in terms of the Froude number, as defined earlier.

From eq. (4.41) for the critical condition

$$(H_s)_{min} = H_c + \frac{v_c^2}{2g}$$

Replacing the last term of this equation by its equivalent from eq. (4.48)

$$(H_s)_{min} = H_c + \frac{H_c}{2} = \frac{3H_c}{2}$$

or

$$H_c = \tfrac{2}{3}(H_s)_{min} \qquad (4.50)$$

Similarly, from eq. (4.45), for Q_{max}

$$\frac{d[H_sH^2 - H^3]}{dH} = 2H_sH - 3H^2 = 0$$

and putting $H_s = (H_s)_{min}$, and $H = H_c$, for the critical condition

$$2H_c(H_s)_{min} - 3H_c^2 = 0$$

$$H_c = \tfrac{2}{3}(H_s)_{min} \tag{4.51}$$

which is the same equation as obtained for a minimum specific head. This also proves that the terms *maximum discharge* and *minimum specific head* are synonymous.

The critical depth is a useful criterion in analysing the surface profiles, as shown in the following diagrams.

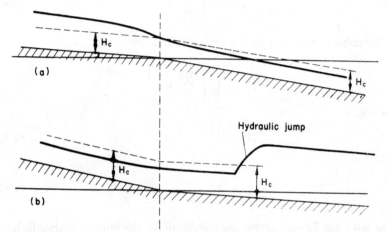

FIG. 4.15. The effect of the slope of the channel's bed on the free surface profile.

Consider, in Fig. 4.15(a), a long channel of a mild slope discharging into one of a steep slope. A uniform subcritical flow will be observed for upstream from the break in the slope while, far downstream from the break, a uniform supercritical flow will be found to exist. The two régimes will be linked by a varied flow in which the transition from subcritical to supercritical flow will gradually take place. In most practical situations, the transition point may be expected to appear in the immediate vicinity of the break in the slope. Since the depth at this point is critical, its measurement provides useful data for the determination of the rate of flow, which can be calculated by means of eq. (4.48).

When the discharge is from a steep channel to a mild one, as in Fig. 4.15(b), uniform flow may be expected upstream and downstream from the point of slope change, as usually the critical depth will not appear in the immediate vicinity of this point. Most commonly, in this situation a hydraulic jump is likely to occur, though its location is difficult to ascertain. The critical depth will then be found within the jump itself.

EXAMPLE 4.10

A rectangular channel, 10 ft wide, is to carry 160 ft³ of water per second. If the roughness factor for the channel is 0·02, what slope will produce critical velocity?

Solution

From eq. (4.46) for critical flow

$$H = H_c = \sqrt[3]{\left(\frac{Q^2}{gb^2}\right)} = \sqrt[3]{\left(\frac{160^2}{32\cdot2\,(10)^2}\right)}$$

$$H_c = 1\cdot995 \text{ ft}$$

Using eq. (4.48)

$$v_c = \sqrt{(gH_c)} = \sqrt{\{(32\cdot2)\,(1\cdot995)\}}$$

$$v_c = 8\cdot02 \text{ fps}$$

The Chèzy coefficient, from eq. (4.11),

$$C = \frac{1\cdot49m^{1/6}}{N} = \frac{1\cdot49m^{1/6}}{0\cdot02}$$

$$C = 74\cdot5m^{1/6}$$

$$Q = CA\sqrt{(mi)} = 74\cdot5Am^{1/6}\sqrt{(mi)}$$

$$v_c = \frac{Q}{A} = 74\cdot5m^{2/3}\sqrt{i}$$

$$A = bH_c = (10)\,(1\cdot995) = 19\cdot95 \text{ ft}^2$$

$$P = b+2H_c = 10+2\,(1\cdot995) = 13\cdot99 \text{ ft}$$

$$m^{2/3} = (A/P)^{2/3} = (19\cdot95/13\cdot99)^{2/3}$$

$$m^{2/3} = 1\cdot267$$

$$8\cdot02 = (74\cdot5)\,(1\cdot267)\sqrt{i}$$

$$i = 0\cdot00722$$

The critical slope is therefore 1 in 138·5.

4.13. Energy Loss in Hydraulic Jump

A hydraulic jump may occur in any situation whenever a sudden change from supercritical to subcritical flow takes place. Typical examples of hydraulic jump may be found downstream from overflow structures or underflow installations such as spillways and sluice gates, respectively.

The theory, relating to the loss of energy in a hydraulic jump, is based on the assumption that the boundary resistance within the jump is negligible, and that the flow immediately upstream and downstream from the front of the jump is uniform. The loss itself is ascribed to the backflow of the upper layer towards the steep wave front resulting in considerable frictional drag and turbulent mixing in the region of contact with the lower layer, moving fast in the direction of channel flow. On the basis of these assumptions, the loss may be calculated from the difference between the specific energy of the stream before and after the jump. Let subscripts 1 and 2 refer to the upstream and downstream conditions respectively, then using eq. (4.41)

$$\Delta H_s = H_{s1} - H_{s2} = \left(H_1 + \frac{v_1^2}{2g}\right) - \left(H_2 + \frac{v_2^2}{2g}\right) \tag{4.52}$$

where ΔH_s is the difference between the specific energies, i.e. the loss of energy per pound of the channel liquid.

The downstream depth is often difficult to measure. It is therefore convenient to express eq. (4.52) in terms of the upstream depth. This can be done by application of the momentum principle, as follows.

Consider, in Fig. 4.16, a horizontal channel carrying a liquid at a supercritical velocity with the effect that a hydraulic jump occurs. Let the subscripts relating to the two sections of the channel considered be as shown in the diagram, then from the principle of momentum

$$(p_2 A_2 - p_1 A_1)g_c = \varrho Q(v_1 - v_2) \tag{4.53}$$

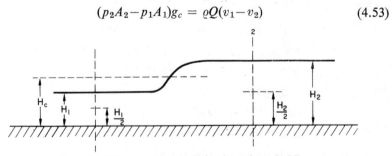

FIG. 4.16. Diagram for the derivation of eq. (4.55).

where p_1 and p_2 are the mean gauge pressures in the centres of gravity of the liquid at the respective sections, as defined by

$$\frac{p_2}{\gamma} = \frac{H_2}{2}$$

$$\frac{p_1}{\gamma} = \frac{H_1}{2}$$

Taking $\gamma = \varrho(g/g_c)$, and substituting from the above equations for

$$p_2 = \frac{H_2}{2}\, \varrho(g/g_c)$$

$$p_1 = \frac{H_1}{2}\, \varrho(g/g_c)$$

in eq. (4.53)

$$\varrho\left(\frac{g}{g_c}\right)\left(\frac{H_2 A_2}{2} - \frac{H_1 A_1}{2}\right) g_c = \varrho Q(v_1 - v_2)$$

Simplifying

$$g(H_2 A_2 - H_1 A_1) = 2Q(v_1 - v_2) \tag{4.54}$$

Assuming steady flow in a rectangular channel of width (b), the following substitutions can be made in eq. (4.54):

$$A_1 = bH_1$$
$$A_2 = bH_2$$
$$v_1 = Q/bH_1$$
$$v_2 = Q/bH_2$$
$$gb\,(H_2^2 - H_1^2) = \frac{2Q^2}{b}\left(\frac{1}{H_1} - \frac{1}{H_2}\right)$$

from which

$$H_2^2 - H_1^2 = \frac{2Q^2}{gb^2}\left(\frac{H_2 - H_1}{H_1 H_2}\right)$$

Simplifying

$$H_2 + H_1 = \left(\frac{2}{g}\right)\left(\frac{Q}{b}\right)^2\left(\frac{1}{H_1 H_2}\right)$$

But

$$Q = bH_1 v_1 = bH_2 v_2$$

hence

$$\left(\frac{Q}{b}\right)^2 = (H_1 v_1)^2$$

and

$$H_2 + H_1 = \frac{2H_1 v_1^2}{g H_2}$$

or

$$H_2^2 + H_1 H_2 - \frac{2H_1 v_1^2}{g} = 0$$

Solving for H_2, and remembering that the depth must be positive

$$H_2 = -\frac{H_1}{2} + \sqrt{\left(\frac{H_1^2}{4} + \frac{2H_1 v_1^2}{g}\right)} \tag{4.55}$$

EXAMPLE 4.11

Water is discharged into a short horizontal channel of rectangular section, at a depth of 1 ft. If the channel is 12 ft wide, what horsepower is lost in the hydraulic jump which occurs in the channel, for a flow of 240 cfs?

Solution

Referring to Fig. 4.16, for $H_1 = 1$ ft,

$$v_1 = \frac{240}{(12)(1)} = 20 \text{ fps}$$

Using eq. (4.55)

$$H_2 = -\frac{1}{2} + \sqrt{\left(\frac{1}{4} + \frac{(2)(1)(400)}{32 \cdot 2}\right)}$$

$$H_2 = 4 \cdot 5 \text{ ft}$$

$$v_2 = \frac{Q}{b H_2} = \frac{240}{(12)(4 \cdot 5)} = \frac{20}{4 \cdot 5}$$

From eq. (4.52)

$$\Delta H_s = (H_1 - H_2) + \frac{v_1^2 - v_2^2}{2g}$$

Substituting the data

$$\Delta H_s = (1 - 4 \cdot 5) + \frac{20^2 - (20/4 \cdot 5)^2}{64 \cdot 4}$$

$$\Delta H_s = 2 \cdot 4 \text{ ft}$$

Taking 62·4 lb/ft³ for the density of water, the total energy lost in the jump is $(\Delta H_s)(Q)(\gamma)$, and equal to

$$(2\cdot4)\,(240)\,(62\cdot4) = 35{,}940\,\frac{\text{(ft) (lb-f)}}{\text{sec}}$$

This is equivalent to

$$\frac{35{,}940}{550} = 65\cdot3 \text{ hp}$$

4.14. Broad-crested Weirs

Unlike sharp-crested weirs, broad-crested weirs are characterised by a non-uniform flow over the crest, except for the region where the flow is reasonably uniform and under critical conditions. These conditions provide the basis for the derivation of a flow equation, which can be obtained as follows.

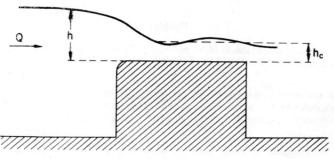

FIG. 4.17. Diagram for the derivation of an equation for a broad-crested weir.

Let, in Fig. 4.17, h_c be the critical depth over the weir and h be the depth measured just upstream from the edge of the crest, where the conditions are reasonably stable, then the flow over the weir may be assumed to be under the differential head $h - h_c$. Hence, for a rectangular section of width b (from eq. (1.42))

$$Q = C_D(bh_c)\sqrt{\{2g(h - h_c)\}}$$
$$Q = C_D b\sqrt{(2g)}\sqrt{(hh_c^2 - h_c^3)} \tag{4.56}$$

where C_D is the coefficient of discharge.

Since, under critical conditions, the flow over the weir is at a maximum, then

$$\frac{dQ}{dh_c} = \frac{1}{2}\,C_D b\sqrt{(2g)}\,(2hh_c - 3h_c^2)\,(hh_c^2 - h_c^3)^{-1/2} = 0$$

$$2hh_c - 3h_c^2 = 0$$

$$h_c = \tfrac{2}{3}h \tag{4.57}$$

Substituting for h_c from this equation in eq. (4.56)

$$Q = C_D b \sqrt{(2g)} \left(\tfrac{2}{3}h\right) \sqrt{\left(h - \tfrac{2}{3}h\right)}$$

$$Q = \frac{2}{3} C_D b \sqrt{(2g)} \sqrt{\left(\frac{h^3}{3}\right)} \tag{4.58}$$

Taking $2g = 64 \cdot 4 \text{ ft/sec}^2$

$$\frac{2}{3} \sqrt{\left(\frac{2g}{3}\right)} \cong 3 \cdot 09$$

and

$$Q = 3 \cdot 09 C_D b h^{3/2} \tag{4.59}$$

For greater accuracy, eq. (4.59) can be corrected for the velocity of approach. Let this velocity be v_0, then $h_0 = v_0^2/2g$, and

$$Q = 3 \cdot 09 \ C_D b (h + h_0)^{3/2} \tag{4.60}$$

Problems

4.1. A rectangular channel, 10 ft wide and laid on a slope of 1 in 1000, conveys water at a depth of 5 ft. If the Chèzy coefficient is 100 $\text{ft}^{1/2}$/sec, what is the flow?

(*Ans.*: 250 cfs)

4.2. A trapezoidal channel, having side walls inclined at 45° to its bottom, conveys water at a depth of 3 ft. If the discharge is 60 cfs at an average velocity of 2·5 fps, what is the width of the channel's bottom and its slope? The Chèzy coefficient is 110 $\text{ft}^{1/2}$/sec.

(*Ans.*: 5 ft, 1 in 3440)

4.3. A trapezoidal channel, laid on a slope of 1 in 6400, has side walls inclined at 45° to its bottom which is 9 ft wide. If the Chèzy coefficient is 80 $\text{ft}^{1/2}$/sec, and the depth of flow is twice the hydraulic radius, what is the flow? (*Ans.*: 503 cfs)

4.4. A trapezoidal channel, having 9 ft wide bottom and side walls inclined to it at 45°, conveys water at a rate of 30,200 lb/sec. If the Chèzy coefficient is 90 $\text{ft}^{1/2}$/sec, and the depth of flow is twice the hydraulic radius, what is the depth, the flow area, and the slope of the channel's bed? The density of water is 62·4 lb/ft^3.

(*Ans.*: 10·86 ft, 215·9 ft^2, 1 in 8750)

4.5. A channel has vertical walls, 6 ft apart, and a semicircular invert laid on a slope of 1 in 3600. What is the value of the Chèzy coefficient if the channel conveys 24 ft^3 of water per second at a centre depth of 4 ft? (*Ans.*: 54 $\text{ft}^{1/2}$/sec.)

4.6. A channel of circular section, having a radius of 12 in., conveys water at a slope of 1 in 3600. If the Chèzy coefficient is 120 $\text{ft}^{1/2}$/sec, what is the flow at a centre depth of 18 in.? (*Ans.*: 3·93 cfs)

4.7. A circular sewer is required to discharge 233·7 cfs of water on a slope of 1 in 3600. If the Chèzy coefficient is given by the equation

$$C = 90 \ m^{1/6}$$

where m is the hydraulic radius, what will be the most economic radius of the circular section? (*Ans.*: 5 ft)

4.8. A sharp-crested weir spans a rectangular channel, 10 ft wide and laid on a slope of 1 in 1000. If the depth measured in the channel just upstream from the weir is 5 ft, what is the depth over the weir itself? The Chèzy coefficient for the channel is 105 ft$^{1/2}$/sec, and the velocity of approach and side contractions may be ignored. (*Ans.*: 3·96 ft)

4.9. A sharp-crested weir spans a rectangular channel, 5 ft wide. The crest of the weir is 2 ft above the bottom of the channel, and the depth measured above the crest is 3 ft. What is the flow if the velocity of approach is ignored, and what is it when allowance is made for this velocity? The side contractions are negligible.
(*Ans.*: 86·6 cfs, 94·8 cfs)

4.10. A rectangular channel, 4 ft wide, is laid on a slope of 1 in 10,000. If the Chèzy coefficient is 100 ft$^{1/2}$/sec, what is the flow at a depth of 2 ft? The flow is measured by a sharp-crested weir spanning the channel. What is the depth above the crest of the weir? The velocity of approach and side contractions may be ignored. (*Ans.*: 8 cfs, 0·71 ft)

4.11. A 90° V-notch measures the flow in a rectangular channel, 3 ft wide. If the depth of flow in the channel is 2 ft, and the measured depth above the apex of the notch is 1 ft, what is the slope of the channel? The Chèzy coefficient is 100 ft$^{1/2}$/sec.
(*Ans.*: 1 in 47,100)

4.12. A rectangular notch, 4 ft wide, is used for flow measurement in a channel. What is the flow, if the measured depth is 18 in.?

If the notch was replaced by a 90° V-notch, with its apex resting on the crest of the original notch, what would be the measured depth for the same flow? The side contractions and velocity of approach may be ignored.
(*Ans.*: 24·5 cfs, 2·47 ft)

4.13. A rectangular channel, 40 ft wide and laid on a slope of 1 in 1600, carries water at a depth of 5 ft. If the Chèzy coefficient is 100 ft$^{1/2}$/sec, what is the flow?

A dam placed across the channel raises the level at the dam to 10 ft. If the value of the Chèzy coefficient remains unchanged, what is the depth 1000 ft upstream from the dam? (*Ans.*: 1000 cfs, 9·46 ft)

4.14. A rectangular channel, 12 ft wide, carries 240 cfs of water at a depth of 4 ft. What is the specific head corresponding to this flow? What would be the critical depth, critical velocity, and minimum specific head for the same flow?
(*Ans.*: 4·388 ft, 2·316 ft, 8·636 fps, 3·474 ft)

4.15. Water flows in a rectangular channel, 12 ft wide, at a depth of 2 ft and a velocity of 10 fps. If the jump occurs, what is the depth after the jump and the horsepower lost in it? (*Ans.*: 2·664 ft, 0·38 hp)

4.16. The depth measured over a broad-crested weir, 20 ft wide, is 4 ft when the flow is 474·6 cfs. What is the coefficient of discharge for the weir? The velocity of approach may be ignored. (*Ans.*: 0·96)

CHAPTER 5

VARIABLE FLOW

UNSTEADY flow which varies linearly with time finds many applications in laboratory and industrial practice. Discharge through orifices or short pipes in emptying tanks, and flow through pipes which have been perforated or fitted with a number of spray nozzles, are typical examples of this kind of flow.

5.1. Flow through Perforated Pipes

Consider a uniformly perforated pipe discharging under a constant head, as shown in Fig. 5.1. Let v_x be the velocity in the pipe at a distance x from the inflow end, then the loss of head in a differential length dx is

$$dh_f = 4f \left(\frac{dx}{D}\right) \left(\frac{v_x^2}{2g}\right) \tag{5.1}$$

where D is the pipe diameter, f is the Darcy friction factor, and v_x is some function of x. The use of this equation is shown in the following example.

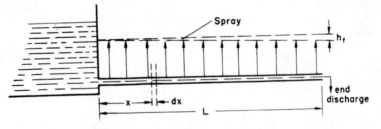

FIG. 5.1. Flow through a perforated pipe.

164

EXAMPLE 5.1

Water is discharged from a reservoir through a pipe, 300 ft long, branching into two pipes of 200 ft length each. All the pipes are of 2 in. diameter and horizontal. The branched pipes are uniformly perforated, one of them being blocked at the end, while in the other one-half the water entering it is drawn off through its end.

If the water surface in the reservoir is maintained at 15 ft above the centre lines of the pipes, what is the discharge through each of the branched pipes? The Darcy friction factor is 0·006.

Solution

Referring to Fig. 5.2, let Q, Q_1, and Q_2 be the rates of flow through the main, the blocked and partly open branched pipes respectively, then

$$Q = Q_1 + Q_2 \tag{1}$$

Also

$$H = h_f + h_{f1} = h_f + h_{f2} \tag{2}$$

where

$$h_f = \frac{fLQ^2}{10\,D^5} = \frac{(0{\cdot}006)\,(300)Q^2}{10\,(2/12)^5}$$

$$h_f = 1400\,Q^2 \tag{3}$$

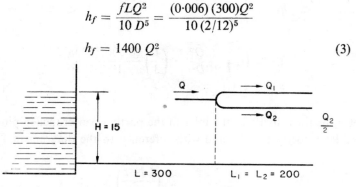

FIG. 5.2. Diagram for Example 5.1.

Let v_x be the velocity at a distance x from the inlet to the blocked pipe, then from the diagram in Fig. 5.3,

$$v_x = v_1\left(\frac{L_1 - x}{L_1}\right)$$

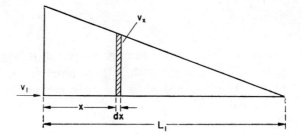

FIG. 5.3. Velocity gradient in the blocked pipe of Example 5.1.

Substituting for v_x in eq. (5.1),

$$dh_{f1} = \frac{4fv_1^2(L_1-x)^2\,(dx)}{2gDL_1^2}$$

$$h_{f1} = \frac{4fv_1^2}{2gDL_1^2} \int\limits_{x=0}^{x=L_1} (L_1-x)^2\,(dx)$$

$$h_{f1} = \frac{4fv_1^2}{2gDL_1^2} \left[L_1^2 x - 2L_1\frac{x^2}{2} + \frac{x^3}{3} \right]_0^{L_1}$$

$$h_{f1} = \frac{1}{3}\,\frac{4fL_1 v_1^2}{2gD}$$

or

$$h_{f1} = \left(\frac{1}{3}\right)\frac{fL_1 Q_1^2}{10D^5} = \left(\frac{1}{3}\right)\frac{(0\cdot006)\,(200)Q_1^2}{10(2/12)^5}$$

$$h_{f1} = 311\cdot1\,Q_1^2 \tag{4}$$

Let v_2 be the velocity at the inlet to the partly open branched pipe, then the outlet velocity is $v_2/2$, and with reference to the diagram in Fig. 5.4

$$v_2 = \frac{v_2}{2}\left(\frac{L_2+z}{z}\right)$$

from which

$$\frac{L_2+z}{z} = 2$$

Also

$$\frac{v_x}{v_2} = \frac{(L_2-x)+z}{L_2+z}$$

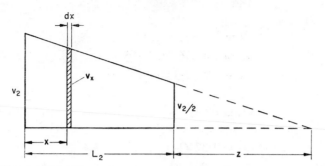

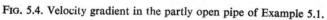

FIG. 5.4. Velocity gradient in the partly open pipe of Example 5.1.

The velocity at a distance x from the branching point is then

$$v_x = v_2 \left(1 - \frac{x}{2L_2}\right)$$

Again, substituting for v_x in eq. (5.1)

$$dh_{f2} = \frac{4fv_2^2 \left(1 - \dfrac{x}{2L_2}\right)^2 (dx)}{2gD}$$

$$h_{f2} = \frac{4fv_2^2}{2gD} \int_{x=0}^{x=L_2} \left(1 - \frac{x}{L_2} + \frac{x^2}{4L_2^2}\right)(dx)$$

$$h_{f2} = \frac{4fv_2^2}{2gD} \left[x - \frac{x^2}{2L_2} + \frac{x^3}{12L_2^2}\right]_0^{L_2}$$

$$h_{f2} = \left(\frac{7}{12}\right) \frac{4fL_2 v_2^2}{2gD}$$

or

$$h_{f2} = \left(\frac{7}{12}\right) \frac{fL_2 Q_2^2}{10D^5} = \left(\frac{7}{12}\right) \frac{(0 \cdot 006)\,(200)Q_2^2}{10(2/12)^5}$$

$$h_{f2} = 554 \cdot 3 Q_2^2 \tag{5}$$

Since $h_{f2} = h_{f1}$, then

$$554 \cdot 3\, Q_2^2 = 311 \cdot 1 Q_1^2$$

$$Q_1 = 1 \cdot 323 Q_2 \tag{6}$$

From eq. (1)

$$Q = Q_1 + Q_2 = 1 \cdot 323 Q_2 + Q_2$$

$$Q = 2 \cdot 323 Q_2 \tag{7}$$

From eq. (2)

$$H = h_f + h_{f2}$$
$$15 = 1400Q^2 + 554 \cdot 3Q_2^2$$
$$15 = 1400(2 \cdot 323)^2 Q_2^2 + 554 \cdot 3Q_2^2$$

from which

$$Q_2 = 0 \cdot 043 \text{ cfs}$$

Also

$$Q_1 = 1 \cdot 323 Q_2 = 1 \cdot 323 (0 \cdot 043)$$
$$Q_1 = 0 \cdot 057 \text{ cfs}$$

5.2. Variable Flow through Orifices and Weirs

Consider, in Fig. 5.5, the discharge from a tank of linearly varying section, through an orifice located at the base of the tank. Let H_1 be the initial head, under which the flow takes place, and H_2 be the head after a time θ.

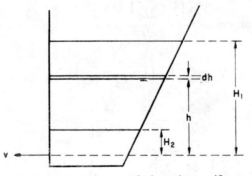

FIG. 5.5. Emptying a tank through an orifice.

Now, assume an instant when the depth measured above the centre of the orifice is h, and let $d\theta$ be the time required to lower this depth by dh, then

$$Q(d\theta) = A(dh) \tag{5.2}$$

where Q is the instantaneous rate of flow through the orifice, and A is the cross-sectional area of the tank at the depth considered.

For a small orifice, the flow is given by eq. (1.42)

$$Q = C_D a \sqrt{(2gh)}$$

where a is the cross-sectional area of the orifice, hence

$$\frac{A(dh)}{d\theta} = C_D a \sqrt{(2gh)}$$

from which

$$d\theta = \frac{Ah^{-1/2}(dh)}{C_D a \sqrt{(2g)}} \tag{5.3}$$

For a uniform section, $A = $ constant, and

$$\theta = \frac{A}{C_D a \sqrt{(2g)}} \int_{h=H_2}^{h=H_1} h^{-1/2}(dh)$$

$$\theta = \frac{2A}{C_D a \sqrt{(2g)}} (H_1^{1/2} - H_2^{1/2}) \tag{5.4}$$

A similar approach can be adopted to variable flow through weirs, as shown in Example 5.4.

EXAMPLE 5.2

A water tank, 12 ft high, forms a frustrum, with diameters of 6 ft and 4 ft at the top and bottom, respectively. If water is discharged through an orifice of 2 in. diameter at the base of the tank, what time will be required to lower the water surface from 9 ft to 4 ft, measured above the orifice? The coefficient of discharge is 0·6.

Solution

Referring to Fig. 5.6, consider a section of the frustrum at a height h above the orifice, then

$$\frac{x}{y} = \frac{h}{12}$$

then, since $y = 1$ ft

$$x = \frac{h}{12}$$

Let D be the diameter at this section, then

$$D = 4 + 2x = 4 + \frac{h}{6}$$

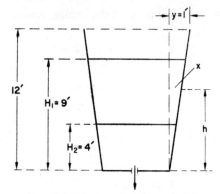

FIG. 5.6. Diagram for Example 5.2.

The cross-sectional area of the section is therefore

$$A = D^2 \frac{\pi}{4} = \left(4 + \frac{h}{6}\right)^2 \frac{\pi}{4}$$

Substituting for A in eq. (5.3), and integrating

$$\theta = \frac{\pi/4}{C_D a \sqrt{(2g)}} \int_{h=4}^{h=9} \left(4 + \frac{h}{6}\right)^2 h^{-1/2}(dh)$$

Evaluating the integral only

$$\int_{h=4}^{h=9} \left(16 + \frac{4}{3} h + \frac{h^2}{36}\right) h^{-1/2} dh = \left[\frac{16h^{1/2}}{0.5} + \frac{8}{9} h^{3/2} + \frac{h^{5/2}}{90}\right]_4^9$$

$$= 51.2$$

$$\theta = \frac{(\pi/4)(51.2)}{(0.6)(2/12)^2 \pi/4 \sqrt{64.4}}$$

$$\theta = 383 \text{ sec.}$$

EXAMPLE 5.3

A cylindrical drum, of 4 ft radius and 12 ft long, is half-full of water. The cylinder lies with its longitudinal axis horizontal. What is the diameter of an orifice, located at the bottom of the cylindrical part of the drum, if emptying the drum takes 50 min 34 sec? The average value of the coefficient of discharge may be assumed to be 0·6.

Solution

Consider, in Fig. 5.7, an instant when the water surface in the drum is h ft above the orifice, then the free surface area at this instant is

$$A = 2xL$$

But

$$x^2 = R^2 - (R-h)^2 = 2Rh - h^2$$

then

$$A = 2L(2Rh - h^2)^{1/2} = 24(8h - h^2)^{1/2}$$

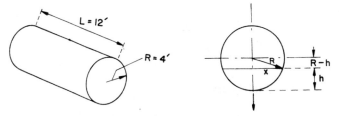

FIG. 5.7. Diagram for Example 5.3.

Substituting the data in eq. (5.3), and taking $a = d^2\pi/4$, when d is the orifice diameter

$$d\theta = \frac{24(8h - h^2)^{1/2} h^{-1/2}(dh)}{(0{\cdot}6)\,(\pi/4)\,\sqrt{64{\cdot}4(d^2)}}$$

$$\frac{24}{(0{\cdot}6)\,(\pi/4)\,\sqrt{64{\cdot}4}} = 6{\cdot}348$$

$$(8h - h^2)^{1/2} h^{-1/2} = (8 - h)^{1/2}$$

$$d\theta = \frac{6{\cdot}348}{d^2}\,(8 - h)^{1/2}\,dh$$

$$\theta = \frac{6{\cdot}348}{d^2} \int_0^4 (8 - h)^{1/2}\,dh = \frac{6{\cdot}348}{d^2}\left[-\frac{2}{3}(8 - h)^{3/2}\right]_0^4$$

$$\theta = \frac{6{\cdot}348}{d^2}\left[\frac{16}{3}\left(\sqrt{(8)} - 1\right)\right]$$

$$\theta = 3034 \text{ sec (given)}$$

$$3034 = \frac{6{\cdot}348}{d^2}\left(\frac{16}{3}\right)(1{\cdot}828)$$

from which

$$d = 0{\cdot}143 \text{ ft } (1{\cdot}72 \text{ in.})$$

EXAMPLE 5.4

A reservoir discharges over a broad-crested weir, 14 ft wide. The water surface covers an area of 300,000 ft² at the level of the weir's crest, and the area increases uniformly at 25,000 ft² per ft height above the crest. If there is no inflow, how long will it take for the water surface to fall from 4 ft to 1 ft above the crest? The coefficient of discharge for the weir is 0·9.

Solution

$$A = 300,000 + 25,000h$$
$$A = 25,000(h+12)$$

Using eq. (4.59)

$$Q = 3\cdot09C_Dbh^{3/2}$$

where h is the height of the water surface above the crest. Substituting the data in eq. (5.2)

$$(3\cdot09)\,(0\cdot9)\,(14)h^{3/2}\,(d\theta) = 25,000(h+12)\,(dh)$$

from which

$$d\theta = \frac{25,000(h+12)\,(dh)}{(3\cdot09)\,(0\cdot9)\,(14)h^{3/2}}$$

$$\theta = \frac{25,000}{(3\cdot09)\,(0\cdot9)\,(14)} \int\limits_{h=1}^{h=4} (h^{-1/2} + 12h^{-3/2})\,(dh)$$

$$\theta = \frac{25,000}{(3\cdot09)\,(0\cdot9)\,(14)} [2h^{1/2} - 24h^{-1/2}]_1^4$$

$$\theta = 8990 \text{ sec}$$

5.3. Variable Flow through a Submerged Orifice

Consider, in Fig. 5.8, a tank of uniform cross-section, partitioned by a vertical wall into two compartments, a smaller one of cross-sectional area A and a larger one of cross-sectional area nA. Assume that the flow takes place between the compartments through a submerged orifice at the base of the partition, and let h be an instantaneous differential head. If Q is the flow at this instant, then in a differential time $d\theta$, the fall of the liquid sur-

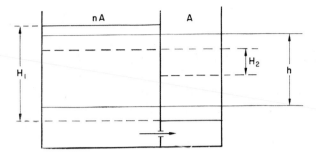

FIG. 5.8. Flow through a submerged orifice.

face in the larger compartment will be $(Q/nA)\,d\theta$. At the same time, the liquid surface in the smaller compartment will rise by $(Q/A)\,d\theta$. The rate of change of the differential head will therefore be

$$\frac{dh}{d\theta} = -\left(\frac{Q}{nA} + \frac{Q}{A}\right) = -\frac{Q(n+1)}{nA}$$

from which

$$Q = -\frac{nA}{1+n}\left(\frac{dh}{d\theta}\right)$$

At the same instant, the rate of flow through the orifice will be (from eq. (1.42))

$$Q = C_D a \sqrt{(2gh)}$$

It follows that

$$\frac{dh}{d\theta} = -C_D a \sqrt{(2gh)}\left(\frac{1+n}{nA}\right)$$

or

$$d\theta = -\frac{nA\,(dh)}{(1+n)C_D a \sqrt{(2gh)}}$$

Integrating between the limits $h = H_2$ and $h = H_1$

$$\theta = -\frac{nA}{(1+n)C_D a \sqrt{(2g)}}\int_{H_1}^{H_2} h^{-1/2}\,(dh)$$

$$\theta = \frac{nA}{(1+n)C_D a \sqrt{(2g)}}\int_{H_2}^{H_1} h^{-1/2}\,(dh) \qquad (5.5)$$

EXAMPLE 5.5

Water is discharged from a large tank to a smaller one through a short pipe, as shown in Fig. 5.9. The cross-sectional areas of the tanks are 4 ft² and 1 ft², respectively. What time will be required to bring the water surfaces in the tanks to the same level, if initially the water surface in the larger tank is 9 ft above that in the smaller one? The pipe may be regarded as being equivalent to an orifice of 1 in. diameter and with a coefficient of discharge of 0·5.

Solution

Referring to Fig. 5·9

$$H_1 = 9 \text{ ft}$$
$$H_2 = 0$$
$$n = 4$$

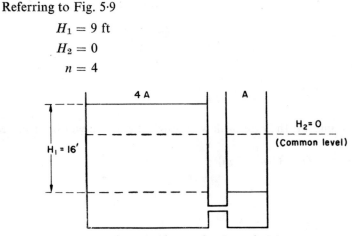

FIG. 5.9. Diagram for Example 5.5.

The value of the integral of eq. (5.5) is

$$\int_{H_2=0}^{H_1=9} h^{-1/2} (dh) = [2h^{1/2}]_0^9 = 6$$

Substituting this value in eq. (5.5)

$$\theta = \frac{(4)(1)(6)}{(1+4)(0·5)(1/12)^2\pi/4\sqrt{64·4}}$$

$$\theta = 220 \text{ sec}$$

5.4. Emptying Tanks through Pipes

Consider, in Fig. 5.10, the discharge through a pipe in the process of emptying a tank. Let h be an instantaneous head under which the flow occurs, then at any instant

$$h = H_f$$

where H_f is the loss of head, mainly in pipe friction. The use of this equation is shown in the following example.

EXAMPLE 5.6

Water is discharged from a cylindrical tank, of 16 ft diameter, through a pipe of 4 in. diameter and 42 ft long, into a pond, as shown in Fig. 5.10. How long will it take to discharge 1500 ft³ of the water, if the water surface in the tank is initially 26 ft above the water surface in the pond, assumed constant? The Darcy friction factor is 0·008.

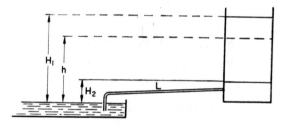

FIG. 5.10. Diagram for Example 5.6.

Solution

The discharge of 1500 ft³ of water lowers the head by

$$\frac{1500}{(16)^2 \, (\pi/4)} = 7\!\cdot\!44 \text{ ft}$$

hence

$$H_2 = 26 - 7\!\cdot\!44 = 18\!\cdot\!56 \text{ ft}$$

Let v be the velocity in the pipe, then allowing for all losses

$$H_f = 0{\cdot}5\,\frac{v^2}{2g} + \frac{v^2}{2g} + 4f\left(\frac{L}{D}\right)\left(\frac{v^2}{2g}\right)$$

$$H_f = \frac{v^2}{2g}\left[0{\cdot}5 + 1{\cdot}0 + 4\,(0{\cdot}008)\left(\frac{42}{4/12}\right)\right]$$

$$h = H_f = 5{\cdot}52\,\frac{v^2}{2g}$$

Putting $2g = 64{\cdot}4$

$$v = 3{\cdot}41 h^{1/2}$$

Let D be the pipe diameter, then

$$Q = \pi/4 D^2 v = (\pi/4)(4/12)^2(3{\cdot}41)h^{1/2}$$

Now let A be the cross-sectional area of the tank, then in a differential time $d\theta$

$$Q(d\theta) = A(dh) \tag{5.2}$$

$$(\pi/4)\,(4/12)^2\,(3{\cdot}41)\,h^{1/2}\,(d\theta) = (16)^2(\pi/4)\,(dh)$$

from which

$$d\theta = 675\,h^{-1/2}\,(dh)$$

$$\theta = 675\int_{H_2}^{H_1} h^{-1/2}\,(dh)$$

$$\theta = (2)\,(675)\,[26^{1/2} - 18{\cdot}56^{1/2}]$$

$$\theta = 1080\ \text{sec}$$

5.5 Discharge with Inflow

The discharge from a tank through an orifice, at a steady inflow, is shown diagrammatically in Fig. 5.11. A fall or rise of the liquid surface in the tank will be observed, depending on whether the inflow is less or more than the initial rate of discharge through the orifice. In each case, however, a point is reached when the inflow equals the outflow, and the surface level from this instant remains stationary.

With reference to Fig. 5.11, let Q_i and Q_o be the constant rate of inflow and the rate of discharge through the orifice at an instant when the head is h, respectively, then at an interval of time $d\theta$

$$Q_i\,(d\theta) = \text{inflow}$$

$$Q_o\,(d\theta) = \text{outflow}$$

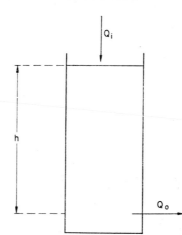

FIG. 5.11. Discharge through an orifice with inflow.

Now, let dh be the change in the surface level, at this interval of time, then

$$(Q_i - Q_o)\,(d\theta) = \pm A\,(dh) \tag{5.6}$$

where A is the cross-section of the tank, and the plus or minus sign indicates the rise and fall of the surface, respectively.

EXAMPLE 5.7

A tank, of 36 ft² cross-sectional area, discharges water through an orifice of 9 in² area, at a steady inflow of 1·2 cfs. How long will it take to raise the water surface from 4 ft to 9 ft above the centre of the orifice? What is the maximum possible height under these conditions? The coefficient of discharge through the orifice is 0·6.

Solution

The discharge through the orifice is (eq. (1·42))

$$Q_o = C_D\,a\,\sqrt{(2gh)} = (0\cdot 6)\,(9/144)\,\sqrt{(64\cdot 4)}\;h^{1/2}$$
$$Q_o = 0\cdot 30\;h^{1/2}$$

Substituting this value and the data in eq. (5·6)

$$(1\cdot 2 - 0\cdot 3\;h^{1/2})\,(d\theta) = 36\,(dh)$$

from which

$$\theta = 36 \int_{h=4}^{h=9} \frac{dh}{1 \cdot 2 - 0 \cdot 3 h^{1/2}}$$

Let $z = 1 \cdot 2 - 0 \cdot 3\, h^{1/2}$, then

$$h = \frac{(1 \cdot 2 - z)^2}{0 \cdot 09}$$

and

$$dh = \frac{-2\,(1 \cdot 2 - z)\,(dz)}{0 \cdot 09} = \frac{(z - 1 \cdot 2)\,(dz)}{0 \cdot 045}$$

$$\theta = \frac{36}{0 \cdot 045} \int \frac{(z - 1 \cdot 2)\,(dz)}{z} = \frac{36}{0 \cdot 045} \int \left(1 - \frac{1 \cdot 2}{z}\right)(dz)$$

$$\theta = \frac{36}{0 \cdot 045} (z - 1 \cdot 2 \log_e z)$$

$$\theta = \frac{36}{0 \cdot 045} [1 \cdot 2 - 0 \cdot 3 h^{1/2} - 1 \cdot 2 \log_e (1 \cdot 2 - 0 \cdot 3\, h^{1/2})]_4^9$$

$$\theta = 423 \text{ sec}$$

At the maximum height, the inflow equals the outflow, hence

$$Q_i = Q_o$$
$$1 \cdot 2 = 0 \cdot 3 H^{1/2}$$
$$H = 16 \text{ ft}$$

EXAMPLE 5.8

A reservoir discharges over a broad-crested weir, 40 ft wide, there being a steady inflow of 60 cfs. The cross-sectional area of the reservoir is 300,000 ft^2 at all levels above the crest of the weir. How long will it take for the water surface to fall from 3 ft to 1 ft above the crest? The coefficient of discharge for the weir is 0·92.

Solution

Let Q_o be the discharge over the weir, then from eq. (4.59)

$$Q_o = 3 \cdot 09\,(0 \cdot 92)\,(40)\,h^{3/2}$$
$$Q_o = 113 \cdot 7\,h^{3/2}$$

where h is the height over the crest of the weir.

Using eq. (5.6)

$$(Q_i - Q_o) (d\theta) = -A(dh)$$

Substituting the data in this equation

$$(60 - 113 \cdot 7 \, h^{3/2}) (d\theta) = -300,000 \, (dh)$$

Integrating

$$\theta = 300,000 \int_{h=1}^{h=3} \frac{dh}{113 \cdot 7h^{3/2} - 60}$$

The evaluation of the integral can be done graphically, or by using Simpson's rule as follows.

Let

$$y = \frac{1}{113 \cdot 7h^{3/2} - 60}$$

then taking h at 0·5 ft intervals

$$h = 1, \quad y_o = \frac{1}{113 \cdot 7 - 60}$$

$$y_o = 0 \cdot 018620$$

Similarly

$$h = 1 \cdot 5, \; y_1 = 0 \cdot 009856$$
$$h = 2 \cdot 0, \; y_2 = 0 \cdot 003824$$
$$h = 2 \cdot 5, \; y_3 = 0 \cdot 002569$$
$$h = 3 \cdot 0, \; y_4 = 0 \cdot 001884$$

By Simpson's rule

$$\int \frac{dh}{113 \cdot 7h^{3/2} - 60} = \frac{\Delta h}{3} (y_0 + 4y_1 + 2y_2 + 4y_3 + y_4)$$

$$= \frac{0 \cdot 5}{3} [0 \cdot 018620 + 4(0 \cdot 009856) + 2(0 \cdot 003824)$$

$$+ 4(0 \cdot 002569) + 0 \cdot 001884)]$$

$$= \frac{0 \cdot 5}{3} (0 \cdot 077852) = 0 \cdot 01298$$

$$\theta = (300,000) (0 \cdot 01298)$$
$$\theta = 3894 \; \text{sec}$$

Problems

5.1. Water flows from a tank through a uniformly perforated pipe, 25 ft long and of 2 in. diameter, its open end being 28·35 ft below the water surface in the tank.

If half of the water entering the pipe is run off the open end (the other half through the perforations), what is the total flow? Losses other than pipe friction may be ignored, and 0·01 may be assumed for the Darcy friction factor. (*Ans.*: 0·5 cfs)

5.2. A uniformly perforated pipe 32 ft long and of 1 in. diameter, is fitted horizontally to a water tap with the other end blocked. If the water (density = 62·4 lb/ft³) is sprayed at a rate of 0·1 cfs, what is the gauge pressure at the tap? The Darcy friction factor is 0·009 and losses other than pipe friction may be ignored. (*Ans.*: 10·35 psig)

5.3. Water flows from a constant head tank through a pipe of 4 in. diameter and 80 ft long. The pipe is uniformly perforated along its length and discharges one-third of the water flowing into it from the tank through the open end. The loss of head may be presented by the equation

$$h_f = \frac{kfLQ^2}{10D^5}$$

where k is a constant. What is the value of the constant under these conditions, assuming no losses other than pipe friction?

If the open end of the pipe is 27 ft below the water surface in the tank, and the Darcy friction factor $f = 0\cdot005$, what is the total flow? (*Ans.*: $k = {}^{13}/_{27}$, 2·4 cfs)

5.4. A vertical tank of 12 ft² cross-sectional area is filled with water to a height of 25 ft, measured above the centre of an orifice located at the side of the tank. If the orifice has a flow area of 0·1 ft², how long will it take to lower the water surface in the tank by 16 ft? The coefficient of discharge for the orifice is 0·6. (*Ans.*: 50 sec)

5.5. A tank, of a uniform cross-section of 1 ft², is filled with water to a depth of 16 ft. The bottom of the tank has an orifice of 1 in² flow area. On opening the orifice, the water surface in the tank falls by 7 ft in 1 min. What is the coefficient of discharge for the orifice? (*Ans.*: 0·6)

5.6. A spherical vessel, of 6 ft radius, contains water to a depth of 4 ft measured above a circular orifice of 6 in. diameter, located at the lowest point of the vessel.

If the coefficient of discharge for the orifice is 0·6, how long will it take to lower the water surface by 3 ft? (*Ans.*: 145 sec)

5.7. A tank, of 16 ft side and square cross-section, has its bottom sloping at 1 to 1. The tank contains water to a depth of 25 ft, measured above an orifice of 0·5 ft² flow area, located at the deepest end of the tank. How long will it take to lower the water surface by 9 ft, and further to drain the tank completely? The coefficient of discharge for the orifice may be assumed to have a value of 0·6. (*Ans.*: 284 sec, 213 sec)

5.8. A water tank has an orifice of 2 ft² flow area located at its rectangular bottom, 50 ft long and 20 ft wide. The side walls of the tank are sloping at 1 to 1, so that they widen upwards. If the tank is initially filled to a depth of 9 ft, how long will it take to drain it? The coefficient of discharge for the orifice is 0·7. (*Ans.*: 794 sec)

5.9. A horizontal boiler, of 3 ft radius and 25 ft long, is half full of water. How long will it take to drain the boiler through a short pipe, equivalent to an orifice of 0·1 ft² flow area and with a coefficient of discharge of 0·6, fitted at the bottom of the boiler? (*Ans.*: 208 sec)

5.10. A vertical plate divides a tank of 48 ft² cross-sectional area into two compartments, resulting in an area ratio of 3 : 1. The compartments are initially filled with water, so that the larger compartment has its water surface 16 ft above that of the smaller one. What time will be required to raise the water level in the smaller compartment by 9 ft, if the flow takes place through a circular orifice of 2 in. diameter, located at the base of the plate and having a coefficient of discharge of 0·6?   (*Ans.*: 457 sec)

5.11. Water flows from a cylindrical tank of 3 ft diameter through a pipe of 2 in. diameter and 100 ft long. How long should it take to lower the water surface in the tank from 9 ft to 1 ft, both measured above the open end of the pipe, if the Darcy friction factor is 0·005? The entry and exit losses to and from the pipe should be allowed for.
(*Ans.*: 594 sec)

5.12. A tank of a uniform cross-section of 160 ft² is drained by a pipe, 40 ft long and of 4 in. diameter. If the liquid surface in the tank is initially 25 ft above the open end of the pipe, by what vertical distance will it be lowered in 4 min? Allowance for entry and exit losses to and from the pipe should be made, and a Darcy friction factor of 0·01 may be assumed.
(*Ans.*: 2·2 ft)

5.13. A tank, of 6 ft diameter, is to be fitted with a 150 ft long pipe. What pipe diameter would be required in order that the water surface in the tank can be lowered from 25 ft to 9 ft, both measured above the open end of the pipe, in 10 min? A Darcy friction factor of 0·01 may be assumed, and losses other than pipe friction may be ignored.
(*Ans.*: 4·22 in.)

5.14. Water flows from a tank of a uniform cross-section of 80 ft² through a circular orifice of 6 in. diameter, at a steady inflow of 1 cfs. If the coefficient of discharge for the orifice is 0·6, how long will it take to lower the water surface in the tank from 16 ft to 9 ft, both measured above the centre of the orifice?

At what level will the water surface remain stationary at this inflow?
(*Ans.*: 462 sec, 1·12 ft)

5.15. A reservoir discharges over a rectangular weir at a steady inflow of 360 cfs. The cross-sectional area of the reservoir is 120,000 ft² at all levels above the crest of the weir, for which the flow equation is given by

$$Q = 40\,h^{3/2}$$

where Q is in cfs, and h is the depth of water measured above the crest of the weir, in ft.

Using Simpson's rule at 0·5 ft intervals, how long will it take to raise the water surface from $h = 2$ ft to $h = 4$ ft?
(*Ans.*: 2057 sec)

CHAPTER 6

CENTRIFUGAL PUMPS

LIKE pumps of the reciprocating type, centrifugal pumps are hydraulic machines for exchanging energy between a mechanical system and a liquid. Their principal object is to generate pressure head.

The essential part of a centrifugal pump is a rotating element, called the *impeller*. It consists sometimes of straight blades, but more commonly it contains curved vanes built into its side walls. These vanes are always curved backwards in relation to the direction of rotation of the impeller, and in the most common type of pump liquid approaches the vanes radially. For this reason pumps of this type are designated *radial*. Large pumps may have doubly curved vanes, but this type will not be considered here.

6.1. Conversion of Velocity Head into Pressure Head

Liquid flows through the impeller outwardly, and enters it with a relatively low velocity and pressure but leaves with both greatly increased due to the centrifugal force acquired in rotation. Since the object of the pump is to raise only its pressure, the high exit velocity would represent a waste of energy imparted to the liquid by the impeller, were it not for the measures taken to convert it into useful pressure energy. The principle involved in this conversion is based simply on making the liquid pass through gradually widening passages where part of its velocity undergoes conversion.

Two distinct approaches have been adopted in design practice to accomplish this energy recovery with the least possible loss.

In one approach, the impeller is surrounded by a chamber of cross-sectional area gradually increasing in the direction of flow. The high velocity with which the liquid leaves the impeller is thus smoothly reduced to the level required in the delivery pipe. The chamber has a spiral form

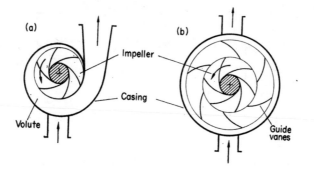

FIG. 6.1. Centrifugal pump. Volute type (a), and turbine type (b).

(Fig. 6.1a) and is called the *volute* or *vortex*. The volute type pump is characterised by low conversion efficiency, usually of less than 50 per cent. Where higher efficiency is essential, as is the case with pumps of greater capacity, a ring of fixed guide-vanes (Fig. 6.1b) is used in place of the volute chamber. Under favourable circumstances, this turbine type pump—as it is often called—is capable of converting some 75 per cent of the velocity into pressure head.

6.2. Velocity Diagrams

Consider, in Fig. 6.2a, a single impeller vane, and let v_0 and v be the respective absolute velocities at which liquid enters and leaves the impeller during its rotation. If u_0 and u are the respective peripheral velocities of the vane tips, the velocity vectors at both ends of the vane will be found

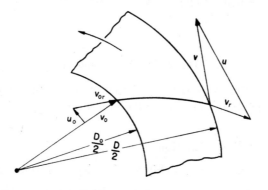

FIG. 6.2a. The velocity triangles.

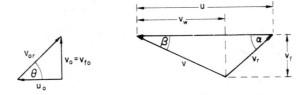

FIG. 6.2b. The velocity triangles redrawn from Fig. 6.2a on a horizontal plane.

to form triangles, the third side of which will be represented by the respective relative velocities, v_{r0} and v_r. The velocity triangles are usually presented with their peripheral velocity components drawn horizontally, as shown in Fig. 6.2(b).

The use of velocity diagrams in calculations is based on the assumption that the relative velocities (v_{0r}) and (v_r) are tangents at their respective vane tips. This condition is met with in all correctly designed pumps in order to eliminate the energy loss due to shock or friction. It is also assumed that the liquid enters the impeller radially. This implies that the inlet velocity vector, $v_0 = v_{f0}$, is radial.

The projection of the absolute outlet velocity (v) on the base of the triangle produces a tangential component v_w. This is the basic velocity component required in the calculation of work done by the impeller. It is known as the *velocity of the whirl*. It may be observed that, in radial flow, the equivalent velocity component at inlet is zero.

The projections of the absolute velocities on the vertical to the base produce the components v_{f0} and v_f, at inlet and outlet triangles respectively, the former being numerically equal to v_0. These velocity components are known as the *velocities of flow*.

6.3. Theoretical Work and Head

The theoretical work done by the impeller on a liquid may be obtained from the gain in energy in passing the liquid through the impeller as follows:

From Newton's law of motion

$$\text{(Force) (Time)} = \text{(Mass) (Velocity)} = \text{Momentum}$$

Let F be the tangential force due to the rotation of the impeller, expressed in the engineering units of lb-f, m be the mass of the liquid leaving the impeller, then on the basis of a unit time of 1 sec

$$Fg_c = mv_w \tag{6.1}$$

where g_c is the gravitational conversion factor to bring the force into the absolute system of units, in compliance with Newton's law, and v_w is the outlet tangential velocity component.

Also taking a unit mass, $m = 1$ lb, and

$$F = \frac{v_w}{g_c}$$

The energy carried by one pound of the liquid as it leaves the impeller is therefore

$$W = Fu = \frac{uv_w}{g_c} \tag{6.2}$$

where u is the tangential distance travelled by the outlet vane tip in 1 sec.

Similarly, the energy carried by 1 lb of the liquid as it enters the impeller is F_0u_0. However, in the radial type pump considered here, $v_{w0} = 0$, then the gain in energy is the outlet energy as given by eq. (6.2).

The relation between theoretical work (W) and head (H) is given from the definition of an engineering unit of work as

$$H = W\frac{g_c}{g} = \frac{uv_w}{g_c}\left(\frac{g_c}{g}\right)$$

It follows that

$$H = \frac{uv_w}{g} \tag{6.3}$$

where g is the gravitational acceleration.

It will be noted that, within the limitation of this text, the two gravitational units have the same numerical value of 32·2.

EXAMPLE 6.1

The impeller of a centrifugal pump has inner and outer diameters of 6 in. and 12 in., respectively, and its vanes are set at an angle of 30° to the outer periphery. If the radial velocity of flow is 5 fps, at both inlet and outlet sides of the impeller, what is the theoretical head produced at

900 rpm, and at what angle are the vane tips inclined to the periphery of the impeller at its inlet side ? If the pump discharges 6 ft³ of water per second, what power is required to drive the shaft?

Solution

The outlet peripheral velocity of the impeller is

$$u = \frac{\pi DN}{60} = \frac{\pi (12/12)(900)}{60}$$

$$u = 47 \cdot 1 \text{ fps}$$

With reference to the diagram of Fig. 6.2b

$$v_w = u - \frac{v_f}{\tan \alpha} = 47 \cdot 1 - \frac{5}{\tan 30} = 47 \cdot 1 - 8 \cdot 65$$

$$v_w = 38 \cdot 45 \text{ fps}$$

Using eq. (6.3)

$$H = \frac{uv_w}{g} = \frac{(38 \cdot 45)(47 \cdot 1)}{32 \cdot 2}$$

$$H = 56 \cdot 2 \text{ ft}$$

The inlet peripheral velocity of the impeller is

$$u_0 = \frac{\pi D_0 N}{60} = \frac{\pi (6/12)(900)}{60}$$

$$u_0 = 23 \cdot 55 \text{ fps}$$

From the inlet triangle (Fig. 6.2)

$$\tan \theta = \frac{v_0}{u_0} = \frac{5}{23 \cdot 35} = 0 \cdot 212$$

$$\theta = 12°$$

The work done by the impeller per pound of water is

$$W = 56 \cdot 2 \text{ (ft) (lb-f/lb)}$$

The mass rate of flow $= Q\varrho = (6)(62 \cdot 4) = 374 \cdot 4$ lb/sec

$$\text{Horsepower} = \frac{WQ\varrho}{550} = \frac{(56 \cdot 2)(374 \cdot 4)}{550} = 38 \cdot 24 \text{ hp}$$

6.4. The Manometric Head

Not all the energy imparted by the impeller to the liquid is utilised in lifting the liquid against the static head, which sustains the flow at the base of the delivery pipe. Some of this energy is lost inside the casing, mainly due to the inefficient conversion of the velocity head produced by the impeller. Let this loss of head be h_m, then

$$H_m = H - h_m \tag{6.4}$$

where H_m is called the *manometric head*, as it can be measured by placing two manometers, one at the base of the delivery pipe and one at the inlet to the impeller. For the same reason, h_m is called the *manometric loss*.

Substituting for H from eq. (6.3) in eq. (6.4)

$$H_m = \frac{uv_w}{g} - h_m \tag{6.5}$$

The manometric loss results mainly from the inefficient conversion of the velocity head, but some additional losses in the casing and in the suction and delivery lines are experienced due to friction. The latter are often of comparatively low magnitude, and may be ignored in calculations.

The manometric loss is more conveniently expressed in terms of a coefficient. The equations (6.4) and (6.5) are then presented in the form

$$H_m = eH \tag{6.6}$$

$$H_m = e\frac{uv_w}{g} \tag{6.7}$$

The name given to the coefficient (e) is the hydraulic or *manometric efficiency* coefficient. Its numerical value depends on the design and operating conditions of the pump, but generally is of the order between 0·5 and 0·75.

EXAMPLE 6.2

The impeller of a centrifugal pump, of an external diameter of 4 ft, has the vanes set at 26° to its external periphery. If the exit radial velocity of flow is 8 fps, what is the manometric efficiency of the pump, running at 200 rpm against a manometric head of 20 ft?

Solution

Referring to the outlet velocity diagram of Fig. 6.2

$$u = \frac{\pi D N}{60} = \frac{(\pi)\,(4)\,(200)}{60}$$

$$u = 41 \cdot 9 \text{ fps}$$

$$v_w = u - \frac{v_f}{\tan \alpha} = 41 \cdot 9 - \frac{8}{\tan 26} = 41 \cdot 9 - 16 \cdot 4$$

$$v_w = 25 \cdot 5 \text{ fps}$$

Using eq. (6.3) the theoretical head produced by the impeller is

$$H = \frac{u v_w}{g} = \frac{(41 \cdot 9)\,(25 \cdot 5)}{32 \cdot 2}$$

$$H = 33 \cdot 2 \text{ ft}$$

At the given manometric head of $H_m = 20$ ft, the manometric efficiency is (from eq. (6.6))

$$e = \frac{H_m}{H} = \frac{20}{33 \cdot 2}$$

$$e = 0 \cdot 6$$

6.5. The Capacity of Centrifugal Pumps

For a given pump, its capacity depends on the radial flow velocity produced by the impeller. The thickness of the vanes reduces the effective flow area, and this has an additional effect on the pump capacity. Let this thickness be t, and n be the number of vanes, then the rate of flow through the impeller, hence the capacity of the pump, is given by

$$Q = (\pi D - nt)\,(b)\,(v_f) = (\pi D_0 - nt)\,(b_0)\,(v_{f0}) \tag{6.8}$$

where (b_0) and (b) are the widths at the respective sides of the impeller.

Equation (6.8) is not necessarily convenient in its application to real problems, and the reduction in the flow area is more commonly expressed on an overall basis, as shown in the following examples.

EXAMPLE 6.3

A centrifugal pump has an impeller of 12 in. external diameter with the vanes set backwards at 35° to its outer periphery. The vanes are 1 in. wide, and their thickness accounts for 4 per cent of the circumferential area. If the pump discharges water at a rate of 30 cfs at 800 rpm, at 0·7 manometric efficiency, what is the manometric head produced?

If the electric motor, which runs the pump, has an efficiency of 95 per cent, and the shaft losses are negligibly small, what power is required to operate the pump?

Solution

Referring to the outer velocity diagram of Fig. 6.2

$$u = \frac{\pi DN}{60} = \frac{(\pi)(12/12)(800)}{60}$$

$$u = 41 \cdot 9 \text{ fps}$$

Allowing 4 per cent for the reduction of circumferential area, the net flow area is

$$A = 0 \cdot 96 \pi Db = 0 \cdot 96 \pi (12/12)(1/12)$$

$$A = 0 \cdot 08 \pi \text{ ft}^2$$

$$v_f = \frac{Q}{A} = \frac{30}{0 \cdot 08 \pi}$$

$$v_f = 11 \cdot 9 \text{ fps}$$

$$v_w = u - \frac{v_f}{\tan \alpha} = 41 \cdot 9 - \frac{11 \cdot 9}{\tan 35}$$

$$v_w = 24 \cdot 9 \text{ fps}$$

$$H = \frac{u v_w}{g} = \frac{(41 \cdot 9)(24 \cdot 9)}{32 \cdot 2}$$

$$H = 32 \cdot 4 \text{ ft}$$

The manometric head produced by the pump at 0·7 efficiency is (from eq. (6.6))

$$H_m = eH = (0 \cdot 7)(32 \cdot 4)$$

$$H_m = 22 \cdot 7 \text{ ft}$$

The work done by the impeller is

$$W = 32 \cdot 4 \, (\text{ft}) \, (\text{lb-f/lb})$$

$$\text{Horsepower} = \frac{WQ\varrho}{(0 \cdot 95)(550)} = \frac{(32 \cdot 4)(30)(62 \cdot 4)}{(0 \cdot 95)(550)}$$

$$118 \cdot 8 \, \text{hp}$$

EXAMPLE 6.4

A centrifugal pump is to discharge 78 cfs at 720 rpm against a manometric head of 60 ft. If the pump is to operate under the following conditions, what should be the external diameter of the impeller and the angle set by the vanes at exit?

$$\begin{aligned} \text{Manometric efficiency} &= 0 \cdot 6 \\ \text{Manometric loss} &= 0 \cdot 016v^2 \\ \text{External flow area} &= 0 \cdot 75D^2 \end{aligned}$$

Solution

$$H = \frac{H_m}{e} = \frac{60}{0 \cdot 6}$$

$$H = 100 \, \text{ft}$$

From eq. (6.4)

$$h_m = H - H_m = 100 - 60$$

$$h_m = 40 \, \text{ft}$$

$$h_m = 0 \cdot 016v^2 \qquad \text{(as given)}$$

$$40 = 0 \cdot 016v^2$$

From which the absolute liquid velocity at exit is

$$v = 50 \, \text{fps} \qquad\qquad (1)$$

Also given $Q = 78$ cfs, and $A = 0 \cdot 75D^2$, and since

$$Q = Av_f$$

then

$$78 = 0 \cdot 75D^2 v_f$$

from which the exit velocity of flow is

$$v_f = \frac{104}{D^2} \tag{2}$$

$$u = \frac{\pi DN}{60} = \frac{720\pi D}{60}$$

$$u = 37 \cdot 68 \ D \ \text{(fps)} \tag{3}$$

From eq. (6.3)

$$v_w = \frac{gH}{u} = \frac{(32 \cdot 2)\,(100)}{37 \cdot 68 D}$$

$$v_w = \frac{85 \cdot 47}{D} \tag{4}$$

Again, referring to the outlet velocity triangle of Fig. 6.2,

$$v_f = \sqrt{(v^2 - v_w^2)}$$

Substituting from the equations (1), (2) and (4)

$$\frac{104}{D^2} = \sqrt{\left(50^2 - \frac{85 \cdot 47^2}{D^2}\right)}$$

Let $x = D^2$

$$\frac{104}{x} = \sqrt{\left(2500 - \frac{7310}{x}\right)}$$

$$2500x^2 - 7310x - 10{,}816 = 0$$

$$x = 4$$

The required diameter of the impeller is then

$$D = 2 \ \text{ft}$$

Substituting for D in the equations (2), (3) and (4)

$$v_f = \frac{104}{4} = 26 \ \text{fps}$$

$$u = (37 \cdot 68)\,(2) = 75 \cdot 36 \ \text{fps}$$

$$u_w = \frac{85 \cdot 47}{2} = 42 \cdot 74 \ \text{fps}$$

From the diagram of Fig. 6.2

$$\tan \alpha = \frac{v_f}{u-v_w} = \frac{26}{75 \cdot 36 - 42 \cdot 74}$$

$$\tan \alpha = 0 \cdot 798$$

The angle at which the vanes are to be set to the exit periphery is then

$$\alpha = 38 \cdot 6°$$

6.6 Pressure Rise through Impeller

The theoretical rise in pressure across the impeller can be calculated from eq. (6.12), which may be derived as follows.

Ignoring the relative position of the vane tips, and assuming no losses in flow between the adjacent vanes, the following energy balance may be set up with the aid of the diagram of Fig. 6.2b.

$$\frac{p}{\gamma} + \frac{v^2}{2g} = \frac{p_0}{\gamma} + \frac{v_0^2}{2g} + \frac{uv_w}{g} \qquad (6.9)$$

where γ is the weight-density of the liquid pumped, and the last term on the extreme side of this equation is the theoretical work done by the impeller. This equation simply states that the energy at outlet equals the energy at the inlet plus the theoretical work done by the impeller.

Let the rise in the pressure head be

$$H_p = \frac{p - p_0}{\gamma}$$

then from eq. (6.9)

$$H_p = \frac{v_0^2}{2g} + \frac{uv_w}{g} - \frac{v^2}{2g} \qquad (6.10)$$

From the outlet velocity triangle (Fig. 6.2b)

$$u - v_w = v_f \cot \alpha$$

from which

$$v_w = u - v_f \cot \alpha \qquad (6.11)$$

Also from the same triangle

$$v^2 = v_f^2 + v_w^2$$

Substituting in this equation from eq. (6.11) for v_w

$$v^2 = v_f^2 + (u - v_f \cot \alpha)^2$$

Expanding this equation, and rearranging the terms

$$v^2 = v_f^2 (1 + \cot^2 \alpha) + u^2 - 2uv_f \cot \alpha$$

Substituting from this equation for v^2, and from eq. (6.11) for v_w, in eq. (6.10)

$$H_p = \frac{v_0^2}{2g} + \frac{u(u - v_f \cot \alpha)}{g} - \frac{v_f^2(1 + \cot^2 \alpha) + u^2 - 2uv_f \cot \alpha}{2g}$$

$$= \frac{v_0^2 + 2u^2 - 2uv_f \cot \alpha - v_f^2 \cosec^2 \alpha - u^2 + 2uv_f \cot \alpha}{2g}$$

Simplifying

$$H_p = \frac{1}{2g}(v_0^2 + u^2 - v_f^2 \cosec^2 \alpha) \tag{6.12}$$

From eq. (6.10) and eq. (6.3)

$$H_p = \frac{v_0^2}{2g} + H - \frac{v^2}{2g}$$

or

$$H = H_p + \frac{v^2 - v_0^2}{2g} \tag{6.12a}$$

This is an alternative equation from which the theoretical head can be evaluated.

EXAMPLE 6.5

The impeller of a centrifugal pump has an external diameter of 24 in., and its vanes are set at 21° to the outer periphery. When running at 600 rpm, the radial velocities are 5 fps at both sides of it. What is the rise in pressure under these conditions? If 65 per cent of the exit velocity head is lost, and in addition 1·7 ft head is lost in friction, what is the manometric efficiency of the pump?

Solution

$$u = \frac{\pi DN}{60} = \frac{\pi(24/12)\,(600)}{60}$$

$$u = 62 \cdot 8 \text{ fps}$$

$$\operatorname{cosec} 21° = 2 \cdot 79$$

Substituting the data in eq. (6.12), and noting that $v_0 = v_f = 5$ fps

$$H_p = \frac{1}{64 \cdot 4}\,(5^2 + 62 \cdot 8^2 - 5^2 \times 2 \cdot 79^2)$$

$$H_p = 58 \cdot 6 \text{ ft}$$

Referring to the outlet triangle of Fig. 6.2b

$$v_w = u - \frac{v_f}{\tan \alpha} = 62 \cdot 8 - \frac{5}{0 \cdot 3839}$$

$$v_w = 49 \cdot 8 \text{ fps}$$

$$v^2 = v_w^2 + v_f^2 = 49 \cdot 8^2 + 5^2$$

$$v^2 = 2505$$

$$\frac{v^2}{2g} = \frac{2505}{64 \cdot 4}$$

$$\frac{v^2}{2g} = 38 \cdot 9 \text{ ft}$$

Velocity head lost

$$(0 \cdot 65)\,(38 \cdot 9) = 25 \cdot 3 \text{ ft}$$

Adding the $1 \cdot 7$ ft due to friction, the manometric loss is

$$h_m = 25 \cdot 3 + 1 \cdot 7 = 27 \cdot 0 \text{ ft}$$

The theoretical head produced by the impeller is (eq. (6.3))

$$H = \frac{u v_w}{g} = \frac{(62 \cdot 8)\,(49 \cdot 8)}{32 \cdot 2}$$

$$H = 97 \cdot 1 \text{ ft}$$

The manometric head produced by the pump is therefore (eq. (6.4))

$$H_m = H - h_m = 97 \cdot 1 - 27 \cdot 0$$

$$H_m = 70 \cdot 1 \text{ ft}$$

From eq. (6.6)

$$e = \frac{H_m}{H} = \frac{70 \cdot 1}{97 \cdot 1}$$

$$e = 0 \cdot 73$$

Alternatively, the theoretical head (H) can be calculated as follows:

$$\frac{v_0^2}{2g} = \frac{25}{64 \cdot 4} = 0 \cdot 26, \quad \text{say} \quad 0 \cdot 3$$

$$\frac{v^2 - v_0^2}{2g} = 38 \cdot 9 - 0 \cdot 3 = 38 \cdot 6$$

Using eq. (6.12a)

$$H = 58 \cdot 6 + 38 \cdot 6 = 97 \cdot 2$$

as compared with 97·1 obtained from eq. (6.3).

6.7. Multi-stage Pumps

For a given pump, the *lift*, as the manometric head is commonly called, depends on the speed of rotation. Any increase in speed, however, also increases losses. At excessive speeds, the losses increase more steeply than the lift, with the effect that the efficiency of the pump starts falling off rapidly. This puts the limit on the working head of a single-impeller pump to about 120 ft.

Where higher heads are in demand, multi-stage pumps are used. These consist of a number of impellers keyed to the same shaft, and housed in a common casing. Liquid flows from one impeller to another in series, and the total head generated is the sum of the heads produced by each of the impellers.

EXAMPLE 6.6

A three-stage centrifugal pump delivers water at a rate of 3·21 cfs, at 800 rpm. The impellers are of the same size, of 18 in. diameter and 2 in. wide at the outlet side, the vanes set being at 45° to the periphery. If the flow area is reduced by 5 per cent by the vanes, and the manometric efficiency of each impeller is 0·75, what is the total manometric head produced?

Solution

$$u = \frac{\pi DN}{60} = \frac{\pi(1 \cdot 5)\,(800)}{60}$$

$$u = 62 \cdot 8 \text{ fps}$$

Allowing 5 per cent for the reduction in the flow area

$$A = 0 \cdot 95 \pi Db = 0 \cdot 95(\pi)\,(1 \cdot 5)\,(2/12)$$

$$A = 0 \cdot 747 \text{ ft}^2$$

The radial velocity of flow at the outlet side of each impeller is

$$v_f = \frac{Q}{A} = \frac{4 \cdot 5}{0 \cdot 747}$$

$$v_f = 6 \cdot 0 \text{ fps}$$

Referring to the outlet triangle of Fig. 6.2b

$$v_w = u - \frac{v_f}{\tan \alpha} = 62 \cdot 8 - \frac{6}{\tan 45}$$

$$v_w = 56 \cdot 8 \text{ fps}$$

The theoretical head developed by each impeller

$$H = \frac{u v_w}{g} = \frac{(62 \cdot 8)\,(56 \cdot 8)}{32 \cdot 2}$$

$$H = 110 \cdot 8 \text{ ft}$$

With 75 per cent manometric efficiency for each impeller, the total mano-metric head developed by the pump is

$$H_m = (3)\,(0 \cdot 75)\,(110 \cdot 8)$$

$$H_m = 249 \text{ ft}$$

6.8. Fans

Fans are used for conveying gases at relatively low pressures. They may be of radial or axial flow type. The design and performance of the radial flow types is based on the same principles as outlined for centri-fugal pumps.

EXAMPLE 6.7

A radial flow type fan conveys air (density = 0·075 lb/ft³) at a rate of 142 cfs, against a manometric head of 70 ft. The impeller has an external diameter of 30 in. and is 4 in. wide, the vanes being set at 60° to the outer periphery. If the impeller rotates at 600 rpm, what is the manometric efficiency of the fan? Assuming 92 per cent efficiency for the driving motor, what horsepower is expended by the fan?

Solution

Neglecting the reduction of the flow area by the vanes

$$A = \pi Db = \pi(30/12)\,(4/12) = \frac{2\cdot5\pi}{3}$$

The radial flow from the impeller

$$v_f = \frac{Q}{A} = \frac{142}{2\cdot5\,\pi/3}$$

$$v_f = 54\cdot3 \text{ fps}$$

With reference to Fig. 6.2b, for the outlet triangle

$$u = \frac{\pi DN}{60} = \frac{\pi(30/12)\,(600)}{60}$$

$$u = 78\cdot5 \text{ fps}$$

$$v_w = u - \frac{v_f}{\tan \alpha} = 78\cdot5 - \frac{54\cdot3}{\tan 60}$$

$$v_w = 47\cdot2 \text{ fps}$$

The theoretical head developed by the impeller is

$$H = \frac{uv_w}{g} = \frac{(78\cdot5)\,(47\cdot2)}{32\cdot2}$$

$$H = 115\cdot1 \text{ ft}$$

At a manometric head of 70 ft the manometric efficiency is

$$e = \frac{H_m}{H} = \frac{70}{115\cdot1}$$

$$e = 0\cdot61$$

The theoretical work done by the impeller on 1 lb of air is

$$W = 115{\cdot}1 \text{ (ft) (lb-f/lb)}$$

The rate of flow of air is

$$(142)\,(0{\cdot}075) = 10{\cdot}65 \text{ lb/sec}$$

At 92 per cent efficiency, the horsepower expended by the fan is

$$\frac{(115{\cdot}1)\,(10{\cdot}65)}{(0{\cdot}92)\,(550)} = 2{\cdot}42 \text{ hp}$$

6.9. Geometrical and Dynamical Similarity of Centrifugal Pumps

Two physical systems are said to be similar, in a general sense, when they show geometrical and dynamical similarity. For such systems, knowledge of the behaviour of one of them enables us to deduce the behaviour of the other.

When applied to centrifugal pumps, the principle of similarity states that two pumps are similar when their respective linear dimensions and velocity diagrams are in a fixed ratio. The usual reference dimension for a centrifugal pump is the diameter of its impeller, and the reference quantity to which velocities are related is the square root of the head developed by the pump. On this basis, the following useful relationships can be deduced.

For a centrifugal pump, the peripheral velocity is given by

$$u = \frac{\pi DN}{60}$$

It follows that

$$u \propto DN$$

or

$$D \propto \frac{u}{N} \tag{6.13}$$

Since the reference quantity to which velocities are related is the square root of the head generated by the impeller, then assuming the theoretical head

$$u \propto \sqrt{H}$$

and substituting for u from this proportionality in eq. (6.13)

$$D \propto \frac{\sqrt{H}}{N} \qquad (6.14)$$

The flow through the impeller is given by

$$Q = \pi Dbv_f$$

and since $v_f \propto \sqrt{H}$, and $b \propto D$, where D is the reference linear dimension, then

$$Q \propto D^2 \sqrt{H} \qquad (6.15)$$

Substituting for D in this equation from eq. (6.14)

$$Q \propto \frac{H\sqrt{H}}{N^2}$$

or

$$Q \propto \frac{H^{3/2}}{N^2} \qquad (6.16)$$

The horsepower expended by a centrifugal pump is related to the flow by

$$P = \frac{Q\rho W}{550}$$

where $W = H(g_c/g)$, then

$$P \propto QH$$

Substituting for Q in this proportionality from eq. (6.16)

$$P \propto \frac{H^{5/2}}{N^2} \qquad (6.17)$$

The principle of similarity can also be adopted to obtain other relationships of interest in pumping. The resulting proportionalities are a useful guide in predicting the characteristics of large-scale pumps from the test data obtained on similar small-scale models.

EXAMPLE 6.8

A model pump has an impeller of 6 in. diameter. When running at 1200 rpm, the pump delivers 2 cfs against a head of 16 ft.

A similar pump is required to discharge 40 cfs at 600 rpm. What should be the diameter of this pump, and what head will it develop?

Solution

Using small letters for the model pump and capitals for the larger pump, then from eq. (6.14)

$$d = k \frac{\sqrt{h}}{n} \qquad\qquad (1)$$

and

$$D = k \frac{\sqrt{H}}{N} \qquad\qquad (2)$$

Substituting for the proportionality constant k from (1) in (2)

$$D = d\left(\frac{n}{N}\right)\frac{\sqrt{H}}{\sqrt{h}} = \left(\frac{6}{12}\right)\left(\frac{1200}{600}\right)\frac{\sqrt{H}}{\sqrt{16}}$$

$$D = 0{\cdot}25\sqrt{H} \qquad\qquad (3)$$

Similarly from eq. (6.15)

$$q = k'd^2\sqrt{h}$$
$$Q = k'D^2\sqrt{H}$$

where k' is another constant.

$$\frac{Q}{q} = \frac{D^2}{d^2}\sqrt{\frac{H}{h}}$$

Putting the data in this equation

$$\frac{40}{2} = \frac{D^2\sqrt{H}}{(6/12)^2\sqrt{16}}$$

$$D^2\sqrt{H} = 20$$

Substituting for $\sqrt{H} = 20/D^2$ from this equation in eq. (3)

$$D = (0{\cdot}25)(20/D^2)$$
$$D^3 = 5$$
$$D = 1{\cdot}71 \text{ ft}$$

Substituting this value in eq. (3)

$$1{\cdot}71 = 0{\cdot}25\sqrt{H}$$

from which

$$H = 46{\cdot}8 \text{ ft}$$

EXAMPLE 6.9

A pump, having an impeller of 12 in. diameter and a flow area of 0·4 ft², with the vanes set at 45° to the periphery, is used for the prediction of design data for a larger pump.

If the model pump delivers water at a rate of 9000 lb/min at 600 rpm, what theoretical horsepower will be required to run a similar pump of 4 ft diameter against a head of 99 ft?

Solution

Using small letters for the model pump, and capital letters for the larger pump, then with reference to the velocity triangle of Fig. 6.2b,

$$u = \frac{\pi d n}{60} = \frac{\pi (12/12) \, (600)}{60}$$

$$u = 31 \cdot 4 \text{ fps}$$

The radial velocity of flow

$$v_f = \frac{9000}{(60) \, (62 \cdot 4) \, (0 \cdot 4)}$$

$$v_f = 6 \text{ fps}$$

$$v_w = u - \frac{v_f}{\tan \alpha} = 31 \cdot 4 - \frac{6}{\tan 45}$$

$$v_w = 25 \cdot 4 \text{ fps}$$

The theoretical head developed by the model pump is (from eq. (6.3))

$$h = \frac{u v_w}{g} = \frac{(31 \cdot 4) \, (25 \cdot 4)}{32 \cdot 2}$$

$$h = 24 \cdot 75 \text{ ft}$$

The work done by the impeller is therefore 24·75 (ft)(lb-f/lb), and since the rate of flow is 9000/60 = 150 lb/sec, then the theoretical horsepower expended by the model pump is

$$p = \frac{(24 \cdot 75) \, (150)}{550}$$

$$p = 6 \cdot 75 \text{ hp}$$

From the proportionality (6.17)

$$p = k\frac{h^{5/2}}{n^2}$$

$$P = k\frac{H^{5/2}}{N^2}$$

Eliminating the proportionality constant (k)

$$P = p\left(\frac{H}{h}\right)^{5/2}\left(\frac{n}{N}\right)^2$$

$$P = 6{\cdot}75\left(\frac{99}{24{\cdot}75}\right)^{5/2}\left(\frac{600}{300}\right)^2$$

$$P = 864 \text{ hp}$$

6.10. Specific Speed

The efficiency of a centrifugal pump depends primarily on its design characteristics. For a given design, however, it also depends on the operational conditions. These should therefore be restricted to within a reasonably narrow range of variables. This range can be predicted, to within a liberal degree of accuracy, with the aid of a dimensional number called the *specific speed*. This is defined as the speed at which a geometrically similar pump will discharge one unit volume of water per minute under a unit head. Equation (6.21), from which this speed can be calculated, may be deduced as follows.

Let Q' be the unit volume discharged by a pump per minute, then from the proportionality (6.16)

$$Q' \propto \frac{H^{3/2}}{N^2} \tag{6.18}$$

from which

$$N \propto \frac{H^{3/4}}{\sqrt{Q'}} \tag{6.19}$$

or

$$N = k\frac{H^{3/4}}{\sqrt{Q'}} \tag{6.20}$$

Now let N_s be the specific speed, as defined earlier, then putting $H = 1$ ft, and $Q' = 1$ ft³/min, in eq. (6.20)

$$N_s = k$$

For a geometrically similar pump we can substitute for the proportionality constant k from this equation in eq. (6.20), and obtain

$$N = N_s \frac{H^{3/4}}{\sqrt{Q'}}$$

or

$$N_s = \frac{N\sqrt{Q'}}{H^{3/4}} \tag{6.21}$$

Taking $Q' = 60Q$, where Q is the volumetric rate of discharge expressed in cfs

$$N_s = \frac{N\sqrt{(60Q)}}{H^{3/4}} \tag{6.22}$$

Experimental evidence has shown that pumps may be run quite efficiently under a wide range of specific speeds. If all types of centrifugal pumps manufactured are taken into consideration, this range may spread between 200 and 6000, on the basis of the units adopted in eq. (6.22) (the equivalent figures between 500 and 15,000 being often quoted where the rate of flow is expressed in imperial gallons per minute). The economical range narrows, however, considerably for one type of pump. For the radial flow type, for example, this range falls within 600 and 1000, approximately. This is a very helpful limitation in predicting an economical performance of the radial-flow type pump.

EXAMPLE 6.10

The impeller of a centrifugal pump has an external diameter of 24 in., a flow area of 1·2 ft², and the vanes are set at 20° to its outer periphery. If a geometrically similar pump runs most efficiently at a specific speed of 597, at what speed should the impeller be run to achieve this efficiency, while discharging water at a rate of 9·6 cfs? What theoretical head will be generated at this speed?

Solution

$$u = \frac{\pi DN}{60} = \frac{\pi(24/12)N}{60}$$

$$u = \frac{\pi N}{30} \tag{1}$$

Referring to the outlet triangle of Fig. 6.2b

$$v_f = \frac{Q}{A} = \frac{9 \cdot 6}{1 \cdot 2}$$

$$v_f = 8 \text{ fps}$$

$$v_w = u - \frac{v_f}{\tan \alpha} = \frac{\pi N}{30} - \frac{8}{\tan 20}$$

$$v_w = \frac{\pi N}{30} - 22 \cdot 0 \tag{2}$$

$$H = \frac{u v_w}{g} = \frac{\left(\dfrac{\pi N}{30}\right)\left(\dfrac{\pi N}{30} - 22\right)}{32 \cdot 2} \tag{3}$$

Substituting in eq. (6.22)

$$597 = \frac{N \sqrt{\{(60)\,(9 \cdot 6)\}}\,(32 \cdot 2)^{3/4}}{\left(\dfrac{\pi N}{30}\right)^{3/4}\left(\dfrac{\pi N}{30} - 22\right)^{3/4}}$$

$$597^{4/3} = \frac{N^{4/3}(576)^{2/3}\,(32 \cdot 2)}{\dfrac{\pi N}{30}\left(\dfrac{\pi N}{30} - 22\right)}$$

$$\frac{\pi N}{30} - 4 \cdot 237 N^{1/3} = 22$$

By trial

$$N = 540 \text{ rpm}$$

Substituting this value in (1), (2) and (3)

$$u = 56 \cdot 5 \text{ fps}$$

$$v_w = 56 \cdot 5 - 22 = 34 \cdot 5 \text{ fps}$$

$$H = \frac{(56 \cdot 5)\,(34 \cdot 5)}{32 \cdot 2}$$

$$H = 60 \cdot 6 \text{ ft}$$

Problems

(Unless stated otherwise, water is the fluid handled in the following problems.)

6.1. A theoretical head of 35 ft is produced by the impeller of a centrifugal pump rotating at 570 rpm. If its larger diameter is 15 in., and the vanes are set at 45° to its external periphery, what is the radial velocity of flow? (*Ans.*: 7·0 fps)

6.2. The theoretical head produced by an impeller of 36 in. diameter is 34·7 ft at 240 rpm. If the radial velocity of flow is 5·6 fps, what is the angle which the vanes make at the periphery? All the data refer to the larger diameter of the impeller.   (*Ans.*: 35°)

6.3. A centrifugal pump delivers 8 cfs at 520 rpm. The impeller has a diameter of 24 in., an effective flow area of 0·8 ft², and vanes set at 50° to the periphery. What manometric head is produced by the pump at 0·65 manometric efficiency? If the driving motor has an efficiency of 95 per cent, what horsepower is expended by the pump? (The data refer to the outer periphery of the impeller.)   (*Ans.*: 50·7 ft, 74·5 hp)

6.4. The impeller of a centrifugal pump has an external diameter of 18 in., and is 2 in. wide at the outer periphery. The vanes are set at 45° at this periphery, and their thickness accounts for 8 per cent of the flow area. If the pump delivers 7·22 cfs at 600 rpm against a manometric head of 35·3 ft, what is the manometric efficiency of the pump?   (*Ans.*: 0·645)

6.5. A centrifugal pump is to deliver 9 cfs at 1200 rpm against a manometric head of 72 ft. What diameter of the impeller will meet the requirements under the following conditions,

$$\text{Manometric efficiency} = 0{\cdot}6$$
$$\text{Manometric loss} \quad = 0{\cdot}01v^2$$
$$\text{Flow area} \quad = 0{\cdot}9D^2$$

where D and v are the larger diameter of the impeller and the velocity at which water leaves the impeller, respectively?   (*Ans.*: 10·8 in.)

6.6. A centrifugal pump is to deliver 267 cfm at 600 rpm against a manometric head of 15 ft. If the vanes of the impeller are set at 30° to its outer periphery, and the radial velocity of flow at this periphery is 10 fps, what external diameter and width will meet these requirements at a manometric efficiency of 0·75? The effect of vane thickness on the flow area may be ignored.   (*Ans.*: 13·6 in., 2 in.)

6.7. The radial velocity of flow is 7 fps at both sides of an impeller rotating with a peripheral speed of 44 fps, at its larger circumference. If the vanes are set at 35° to this periphery, what is the theoretical head produced, and the pressure rise across the impeller? If the manometric efficiency is 0·68, and the frictional losses add 1·8 ft to the manometric loss, what percentage of the exit velocity head is recovered?

(*Ans.*: 46·5 ft, 28·5 ft, 30 per cent)

6.8. A four-stage centrifugal pump is to work against a manometric head of 160 ft at 360 rpm.

If the vanes of each impeller are set at 45° to their external periphery, and the radial velocity of flow is

$$v_f = 0{\cdot}17u$$

where u is the external peripheral speed, what impeller diameter will meet the requirements at a manometric efficiency of 0·7?   (*Ans.*: 2·5 ft),

6.9. A fan requires 5 hp to convey air (density = 0·075 lb/ft³) at a rate of 192 cfs, when its impeller rotates at 720 rpm. If the impeller has an external diameter of 3 ft, and the driving motor has an efficiency of 95 per cent, what manometric head does the fan develop at a manometric efficiency of 0·65? If the radial velocity of flow at the outlet periphery of the impeller is 60 fps, what angle do the vanes make at this periphery?   (*Ans.*: 118·9 ft, 45°)

6.10. A model pump delivers 2 cfs at 600 rpm against a head of 20 ft. What should be the speed of rotation of a geometrically similar pump to deliver 18 cfs under a head of 30 ft? What should be the diameter ratio of the two impellers? (*Ans.*: 271 rpm, 2·71)

6.11. A model pump requires 22 hp when running at 600 rpm against a head of 25 ft.

A geometrically similar pump, with a diameter four times larger than that of the model, is required to operate against a head of 100 ft. What should be the speed of rotation for the pump? What horsepower will it require?   (*Ans.*: 300 rpm, 2816 hp)

6.12. The impeller of a centrifugal pump has its larger diameter of 4 ft and the vanes set at 31° to the outer periphery. The pump delivers 2 cfs at a radial velocity of 6 fps, when the external peripheral velocity is 30 fps. What theoretical head is produced by the impeller? At what specific speed is the pump operated?   (*Ans.*: 18·6 ft, 175)

6.13. The impeller of a centrifugal pump has a larger diameter of 18 in., and the vanes set at 45° to the outer periphery. The pump is to deliver 9 cfs at a radial velocity of 8 fps, under a specific speed of 800. What should be the speed of rotation of the impeller? What theoretical head will it develop?   (*Ans.*: 462 rpm, 31·8 ft)

6.14. Assuming that the volumetric discharge (Q) of a centrifugal pump depends on the speed of rotation of the impeller (N), its diameter (D), the theoretical head produced per unit mass of the liquid pumped (H), and on the density of the fluid (ϱ), the following equation can be obtained on the application of the principle of dimensional analysis

$$Q = k(\varrho D^2 H)^t$$

where k and t are some constants. What group of terms does the constant k represent?
     (*Ans.*: $k = ND^3$)

CHAPTER 7

COMPRESSIBLE FLOW

THERE is no sharp demarcation recognised by any convention between compressible and non-compressible flow. From a practical point of view, flow is said to be compressible if pressure changes in the direction of flow affect fluid density appreciably. On this basis, liquids and gases flowing under conditions of moderate pressure changes are excluded from the analysis of this kind of flow.

7.1. Mach Number

The rate of change of fluid density with respect to pressure is an important parameter in analysing compressible flow. It is closely related to the velocity of propagation of small pressure disturbances characteristic of the velocity of sound, more commonly known as *acoustic* or *sonic velocity*. This velocity therefore serves as a convenient basis for the classification of compressible flow.

Let v be the average velocity of flow, and v_s be the sonic velocity in the same fluid medium, then

$$M = \frac{v}{v_s} \qquad (7.1)$$

where M is the symbol usually given to the so-called *Mach number*.

Compressible flow is classified in terms of this number as follows:

Subsonic	$M < 1$
Sonic	$M = 1$
Supersonic	$M > 1$
Hypersonic	$M > 5$

When fluid velocity, relative to a solid boundary, approaches its sonic level, resistance to flow increases abruptly. It reaches its maximum value at the sonic level and shows a gradual decline above this level. The sudden increase in resistance commences at a Mach number of about 0·6, and the range between this number and the Mach number just exceeding unity is of special interest in design practice. This range is called *transonic*. The change in the resistance may be so rapid that a sensation, resembling explosion, may be evidenced within this range. This is commonly referred to as the *sonic bang*.

7.2. Sonic Velocity

Many physical processes follow approximately either an isothermal or an adiabatic pattern, i.e. they proceed with either no temperature changes, or no heat interchanges, respectively. If, in addition to being adiabatic, a process is undergone in a frictionless manner (reversible process), it follows a constant entropy pattern, and is called *isentropic*. The isentropic behaviour of gases is a very useful concept in the analysis of flow phenomena, as it provides a model or limit for a real non-adiabatic process. The concept is also of immense value in the derivation of the basic equation for sonic flow.

Sonic velocity is defined as the speed with which a small pressure disturbance is transmitted through a fluid medium. The algebraic form of the definition is

$$\frac{E}{\varrho} = \frac{dp}{d\varrho} \tag{7.2}$$

where E is the bulk modulus for the fluid (analogous to the Young's modulus of elasticity for solid rods), and ϱ is its density.

The square root of the modulus–density ratio, i.e. the term $\sqrt{(E/\varrho)}$, has the dimension of velocity, and it can be proved[11, 12] that this is the sonic velocity, so that

$$v_s = \sqrt{\frac{E}{\varrho}}$$

or

$$v_s = \sqrt{\frac{dp}{d\varrho}} \tag{7.3}$$

Equation (7.3) shows that the sonic velocity is inversely proportional to the ratio $d\varrho/dp$ which represents the rate of change of density with pressure. This ratio also defines an important property of the fluid, namely its *compressibility*. For an *isentropic expansion*

$$pV^k = C \qquad (7.4)$$

where C and k are constants, the latter being the ratio of the specific heats at constant pressure and constant volume, respectively.

Since the specific volume $V = 1/\varrho$, then from eq. (7.4)

$$p = C\varrho^k \qquad (7.5)$$

Presenting this equation in a differential form

$$\frac{dp}{d\varrho} = kC\varrho^{k-1}$$

Substituting in this equation, from eq. (7.5), for

$$C = \frac{p}{\varrho^k}$$

$$\frac{dp}{d\varrho} = k\left(\frac{p}{\varrho^k}\right)(\varrho^{k-1})$$

Simplifying

$$\frac{dp}{d\varrho} = k\frac{p}{\varrho}$$

or

$$\frac{dp}{d\varrho} = kpV \qquad (7.6)$$

Substituting for $dp/d\varrho$ from this equation in eq. (7.3)

$$v_s = \sqrt{(kpV)} \qquad (7.7)$$

An alternative form of the sonic velocity equation can be obtained by introducing the universal gas constant R (from eq. (1.8)), so that

$$v_s = \sqrt{(kRT/M_w)} \qquad (7.8)$$

EXAMPLE 7.1

Air ($k = 1\cdot4$, $M_w = 29$) flows at a velocity of 224 fps under a pressure of 14·7 psia. If the temperature of the air is 17°C, at what Mach number does the flow take place?

Solution

At 17°C, and normal pressure, the specific volume of air is

$$V = \frac{(359)\,(273+17)}{(29)\,(273)}$$

$$V = 13{\cdot}15 \text{ ft}^3/\text{lb}$$

$$p = (14{\cdot}7)\,(144)(32{\cdot}2) = 68{,}170 \text{ poundal/ft}^2$$

Using eq. (7.7)

$$v_s = \sqrt{\{(1{\cdot}4)\,(68{,}170)\,(13{\cdot}15)\}}$$

$$v_s = 1120 \text{ fps}$$

(It will be noted that the NACA standard sonic velocity of air at 59°F is 1117 fps.)

$$M = \frac{v}{v_s} = \frac{224}{1120}$$

$$M = 0{\cdot}2$$

7.3. Pressure Drop in Long Pipes

Consider a subsonic flow in a long horizontal pipe. Let the flow be adiabatic ($q = 0$) and assume no shaft work ($W = 0$), then eq. (1.34) reduces to

$$\frac{p}{\gamma} + \frac{v^2}{2ag} + H_f = \text{constant} \qquad (7.9)$$

In compressible flow, the Reynolds number is likely to be very high, then the correcting factor α approaches unity, and since

$$\gamma = \varrho \left(\frac{g}{g_c} \right) \qquad (1.6)$$

then

$$\frac{pg_c}{g\varrho} + \frac{v^2}{2g} + H_f = \text{constant}$$

Multiplying all terms of this equation by g/g_c, and recalling that $\varrho = 1/V$

$$pV + \frac{v^2}{2g_c} + \frac{g}{g_c}\,(H_f) = \text{constant} \qquad (7.10)$$

Expressing this equation in a differential form

$$d(pV) + \frac{dv^2}{2g_c} + \frac{g}{g_c}(dH_f) = 0$$

$$V(dp) + \frac{v(dv)}{g_c} + \frac{g}{g_c}(dH_f) = 0 \qquad (7.11)$$

In a straight pipe, losses other than pipe friction may be ignored; then

$$H_f = h_f = 4f\frac{L}{D}\frac{v^2}{2g} \qquad (3.15)$$

and

$$dH_f = \frac{4fv^2(dL)}{2gD}$$

Substituting for dH_f from this equation in eq. (7.11), and simplifying

$$V(dp) + \frac{v(dv)}{g_c} + \frac{4fv^2(dL)}{2g_cD} = 0 \qquad (7.12)$$

Dividing all terms of this equation by V^2

$$\frac{dp}{V} + \frac{v(dv)}{V^2g_c} + \frac{4f}{2g_cD}\left(\frac{v}{V}\right)^2(dL) = 0 \qquad (7.13)$$

Introducing the concept of mass velocity (eq. (1.15))

$$G = v\varrho = \frac{v}{V} = \text{constant}$$

from which

$$v = GV \qquad (7.14)$$

and

$$dv = G(dV)$$

Substituting in eq. (7.13) for v from eq. (7.14) and for

$$v(dv) = G^2V(dV) \qquad (7.15)$$

$$\frac{dp}{V} + \frac{G^2V(dV)}{V^2g_c} + \frac{4fG^2(dL)}{2g_cD} = 0$$

Again, substituting from eq. (1.8) in the first term of this equation for

$$V = \frac{RT}{pM_w}$$

$$\frac{p(dp)}{RT/M_w} + \frac{G^2(dV)}{g_cV} + \frac{4fG^2(dL)}{2g_cD} = 0 \qquad (7.16)$$

Let the subscripts (1) and (2) refer to the inlet and outlet ends of the pipe, then integrating eq. (7.16) between these limits

$$\frac{M_w}{RT} \int_2^1 p(dp) + \frac{G^2}{g_c} \int_2^1 \frac{dV}{V} + \frac{4fG^2}{2g_cD} \int_2^1 (dL) = 0$$

$$\frac{M_w}{RT} \int_2^1 p(dp) = \frac{G^2}{g_c} \int_1^2 \frac{dV}{V} + \frac{4fG^2}{2g_cD} \int_1^2 (dL)$$

$$\frac{p_1^2 - p_2^2}{2RT/M_w} = \frac{G^2}{g_c} \log_e \frac{V_2}{V_1} + \frac{4fLG^2}{2g_cD} \qquad (7.17)$$

Equation (7.17) is sometimes referred to as the Weymouth equation. In the derivation of this equation, the temperature has been assumed constant. It should therefore be taken at an average value.

EXAMPLE 7.2

A natural gas ($M_w = 16$) is at a steady flow to a storage tank through a horizontal pipe of 12 in. diameter, and 16,000 ft long. The pressure in the tank is maintained at 15·7 psia, and the gas in the pipe is at a uniform temperature of 17°C. What horsepower is expended in pumping the gas at a rate of 470·1 lb/min? The Darcy friction factor is 0·012.

Solution

$$G = \frac{470 \cdot 1}{60\pi/4}$$

$$G = 10 \cdot 0 \text{ lb/(ft}^2) \text{ (sec)}$$

$$\frac{4fLG^2}{2g_cD} = \frac{(4)(0 \cdot 012)(16,000)(10^2)}{(2)(32 \cdot 2)(1)} = 1192$$

$$\frac{R}{M_w} = \frac{(14 \cdot 7)(144)(359)}{(273)(16)} = 174 \frac{\text{(lb-f) (ft)}}{\text{(lb) (°C)}}$$

$$\frac{2RT}{M_w} = (2)(174)(273 + 17) = 100,900$$

$$p_2 = (15 \cdot 7)(144) = 2261 \text{ psf}$$

Substituting in eq. (7.17), and ignoring the first right-hand-side term of this equation as negligibly small,

$$\frac{p_1^2-(2261)^2}{100,900} = 1192$$

$$p_1^2 = 5{\cdot}12\times10^6 + 1{\cdot}203\times10^8$$

$$p_1 = 11,200 \text{ psf}$$

This is equivalent to 77·8 psia. Now, it can be shown that the ignored term

$$\frac{G^2}{g_c}\log_e\frac{V_2}{V_1} = \frac{G^2}{g_c}\log_e\frac{p_1}{p_2} = \frac{100}{32{\cdot}2}\log_e\frac{11,200}{2261} \cong 5$$

is very small indeed when compared with the 1192 figure obtained for the last term of eq. (7.17).

The average pressure in the pipe is

$$\frac{p_1+p_2}{2} = \frac{11,200+2261}{2} = 6730 \text{ psf}$$

The mean density of the gas at this pressure is

$$\varrho = \frac{(16)\,(273)\,(6730)}{(359)\,(290)\,(14{\cdot}7)\,(144)}$$

$$\varrho = 0{\cdot}1334 \text{ lb/ft}^3$$

The average volumetric rate of flow

$$Q = \frac{47{\cdot}01/60}{0{\cdot}1334} = 5{\cdot}875 \text{ cfs}$$

$$\text{Horsepower} = \frac{(p_1-p_2)Q}{550} = \frac{(11,200-2261)\,(5{\cdot}875)}{550}$$

$$\text{Horsepower} = 95{\cdot}5$$

7.4. Pressure Drop in Short Pipes

Dividing eq. (7.12) by V, and rearranging the terms,

$$-dp = \frac{v(dv)}{g_c V} + \frac{4fv^2(dL)}{2g_c VD}$$

Substituting in this equation for v and $v(dv)$ from the corresponding equations (7.14) and (7.15), and simplifying,

$$-dp = \frac{G^2(dV)}{g_c} + \frac{4fG^2V(dL)}{2g_cD} \tag{7.18}$$

Equation (7.18) can be easily integrated if the specific volume, which appears in the last term of this equation, is taken as a constant. Let

$$V = \frac{V_1 + V_2}{2} = \text{constant}$$

then

$$-\int_1^2 dp = \frac{G^2}{g_c} \int_1^2 dV + \frac{4fG^2V}{2g_cD} \int_1^2 dL$$

$$p_1 - p_2 = \frac{G^2}{g_c}(V_2 - V_1) + \frac{4fG^2VL}{2g_cD} \tag{7.19}$$

Although less rigorous than eq. (7.17), eq. (7.19) is in common use in engineering problems for its simplicity.

EXAMPLE 7.3

A gas ($M_w = 29$) is cooled inside tubes of 1 in. diameter and 20 ft long, at a mass-velocity of 2·125 lb/(ft²)(sec). If the respective temperatures at the ends of the tubes are 500°F and 180°F, what is the drop in pressure? The Darcy friction factor is 0·0075, and the outlet pressure is normal atmospheric.

Solution

Let the subscripts (1) and (2) refer to the inlet and outlet of the tubes, respectively, then ignoring at this stage a small variation of pressure inside the tubes,

$$V_1 = \frac{(359)(460 + 500)}{(29)(492)}$$

$$V_1 = 24 \cdot 1 \ \text{ft}^3/\text{lb}$$

$$V_2 = (24 \cdot 1)\frac{(460 + 180)}{(460 + 500)}$$

$$V_2 = 16 \cdot 1 \ \text{ft}^3/\text{lb}$$

$$V = \frac{V_1 + V_2}{2} = \frac{24 \cdot 1 + 16 \cdot 1}{2}$$

$$V = 20 \cdot 1 \ \text{ft}^3/\text{lb}$$

Using eq. (7.19)

$$p_1 - p_2 = \frac{(2 \cdot 125)^2 (16 \cdot 1 - 24 \cdot 1)}{32 \cdot 2} + \frac{(4)(0 \cdot 0075)(2 \cdot 125)^2 (20)}{(64 \cdot 4)(1/12)}$$

$$= -1 \cdot 1 + 10 \cdot 1$$

$$p_1 - p_2 = 9 \text{ psf}$$

7.5. Variable-area Flow

Consider, in Fig. 7.1, an isentropic flow through a convergent–divergent type nozzle. From the Bernoulli theorem, using the absolute system of units

$$p + \frac{\varrho v^2}{2} + \varrho Z g = \text{constant} \qquad (1.29)$$

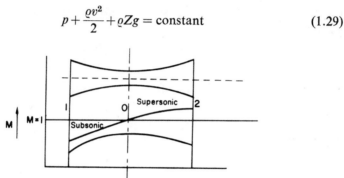

FIG. 7.1. Isentropic flow through convergent–divergent nozzle.

Dividing all terms of this equation by ϱ,

$$\frac{p}{\varrho} + \frac{v^2}{2} + Zg = \text{constant}$$

Writing this equation in a differential form, and noting that for a horizontal nozzle Z is a constant,

$$\frac{dp}{\varrho} + v(dv) = 0 \qquad (7.20)$$

This form was presented in 1750 by Euler, and eq. (7.20) bears his name. The equation may be written

$$\left(\frac{dp}{d\varrho}\right)\left(\frac{d\varrho}{\varrho}\right) + v(dv) = 0$$

But, from eq. (7.3),

$$\frac{dp}{d\varrho} = v_s^2$$

then

$$(v_s^2)\frac{d\varrho}{\varrho} + v(dv) = 0$$

from which

$$\frac{d\varrho}{\varrho} = -\frac{v(dv)}{v_s^2}$$

or

$$\frac{d\varrho}{\varrho} = -\frac{dv}{v}\left(\frac{v}{v_s}\right)^2$$

Introducing at this stage the Mach number (eq. (7.1)), the last equation becomes

$$\frac{d\varrho}{\varrho} = -M^2\left(\frac{dv}{v}\right) \tag{7.21}$$

Let A be the variable flow area in the nozzle, then by the law of continuity

$$Av\varrho = \text{constant}$$

Partial differentiation of this equation gives

$$\frac{dA}{A} + \frac{dv}{v} + \frac{d\varrho}{\varrho} = 0$$

from which

$$\frac{d\varrho}{\varrho} = -\frac{dA}{A} - \frac{dv}{v}$$

Substituting from this equation for $d\varrho/\varrho$ in eq. (7.21)

$$-\frac{dA}{A} - \frac{dv}{v} = -M^2\frac{dv}{v}$$

or

$$\frac{dA}{A} = (M^2 - 1)\frac{dv}{v} \tag{7.22}$$

Inspection of eq. (7.22) shows that in the divergent section of the nozzle, where $dA/A > 0$, the term $(M^2 - 1)\,dv/v$ must be positive; then either $M < 1$ and $dv/v < 0$, or $M > 1$ and $dv/v > 0$. In the latter case, the flow in the divergent section is supersonic, and this condition requires sonic flow in the throat.

Equation (7.22) also shows that the maximum velocity at the throat is the sonic velocity, as at this section $dA/A = 0$, and $dv/v = 0$, so that $M^2 - 1 = 0$ and $M = 1$. If the gas velocity in the throat is subsonic, there is no chance for the gas to attain supersonic velocity in either section of the nozzle. Its velocity increases in the convergent section and decreases in the divergent section, as shown graphically in Fig. 7.1.

7.6. Subsonic Flow in Nozzles

Consider a compressible flow between two tanks through a convergent-type nozzle, as shown in Fig. 7.2. The equation which applies to this situation is obtained from eq. (7.20). Taking $V = 1/\varrho$ in this equation

$$V(dp) + v(dv) = 0$$

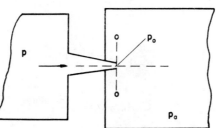

FIG. 7.2. Flow through a convergent nozzle.

Assuming that the expansion through the nozzle follows the law

$$pV^n = C \tag{7.23}$$

or

$$p = CV^{-n}$$

then

$$dp = -nCV^{-n-1}(dV)$$

Substituting from the last equation for (dp) in eq. (7.20) and simplifying

$$-nCV^{-n}(dV) + v(dv) = 0$$

Rearranging and integrating

$$\int_{v_0}^{v} v\,(dv) = nC \int_{V_0}^{V} V^{-n}\,(dV)$$

$$\frac{v^2 - v_0^2}{2} = \frac{nC}{1-n}(V^{1-n} - V_0^{1-n})$$

Upstream from the nozzle the gas is stationary, and $v = 0$, then

$$v_0^2 = -\frac{2nC}{1-n}(V^{1-n} - V_0^{1-n})$$

Changing the sign, and using eq. (7.23)

$$C = pV^n = \frac{pV}{V^{1-n}}$$

$$v_0^2 = \frac{2npV}{n-1}\left[1 - \left(\frac{V_0}{V}\right)^{1-n}\right]$$

But

$$\frac{V_0}{V} = \left(\frac{p}{p_0}\right)^{1/n}$$

and

$$\left(\frac{V_0}{V}\right)^{1-n} = \left(\frac{p}{p_0}\right)^{(1-n)/n} = \left(\frac{p_0}{p}\right)^{(n-1)/n}$$

then

$$v_0^2 = \frac{2npV}{n-1}\left[1 - \left(\frac{p_0}{p}\right)^{(n-1)/n}\right] \tag{7.24}$$

Now let

$$W = C_D A_0 v_0 \varrho_0 = \frac{C_D A_0 v_0}{V_0}$$

where W is the mass rate of flow, and C_D is the coefficient of discharge for the nozzle, then substituting from this equation for v_0 in eq. (7.24)

$$\frac{W^2}{C_D^2 A_0^2} = \frac{2npV}{(n-1)V_0^2}\left[1 - \left(\frac{p_0}{p}\right)^{(n-1)/n}\right]$$

Substituting in this equation for

$$\frac{pV}{V_0^2} = \frac{p}{V}\left(\frac{V}{V_0}\right)^2 = \frac{p}{V}\left(\frac{p_0}{p}\right)^{2/n} = p\varrho\left(\frac{p_0}{p}\right)^{2/n}$$

$$\frac{W^2}{C_D^2 A_0^2} = \frac{2np\varrho}{n-1}\left[\left(\frac{p_0}{p}\right)^{2/n} - \left(\frac{p_0}{p}\right)^{(n+1)/n}\right]$$

from which

$$W = C_D A_0 \sqrt{\left\{\frac{2np\varrho}{n-1}\left[\left(\frac{p_0}{p}\right)^{2/n} - \left(\frac{p_0}{p}\right)^{(n+1)/n}\right]\right\}} \tag{7.25}$$

where p and ϱ are the pressure (in the absolute units) and density of the gas, respectively, upstream from the nozzle, i.e. in the higher pressure tank.

EXAMPLE 7.4

A gas ($M_w = 2$, $n = 1\cdot4$) is at a steady flow from a gas holder to a receiver through a converging type nozzle of 2 in. diameter throat. The pressures in the receiver and the gas holder are 15 psia and 25 psia, respectively, and the temperature in the latter is 18°C. If the coefficient of discharge for the nozzle is 0·98, what is the flow?

Solution

$$p = (25)(144)(32\cdot2)$$

$$p = 115{,}920 \text{ poundal/ft}^2$$

$$\varrho = \frac{(2)(25)(273)}{(359)(14\cdot7)(273+18)}$$

$$\varrho = 0\cdot0089 \text{ lb/ft}^3$$

$$A_\varrho = \left(\frac{2}{12}\right)^2 \frac{\pi}{4}$$

$$A_\varrho = \frac{\pi}{144} \text{ ft}^2$$

The pressure at the throat of the nozzle is the same as that in the receiver, then

$$\left(\frac{p_0}{p}\right)^{2/n} = \left(\frac{15}{25}\right)^{2/1\cdot4} = 0\cdot482$$

$$\left(\frac{p_0}{p}\right)^{(n+1/n)} = \left(\frac{15}{25}\right)^{2\cdot4/1\cdot4} = 0\cdot417$$

Substituting in eq. (7.25)

$$W = (0\cdot98)\frac{\pi}{144}\sqrt{\left(\frac{(2)(1\cdot4)(115{,}920)(0\cdot0089)(0\cdot482-0\cdot417)}{0\cdot4}\right)}$$

$$W = 0\cdot463 \text{ lb/sec}$$

7.7. Critical Pressure Ratio

Consider, with reference to Fig. 7.2, an isentropic flow through a convergent nozzle under a constant upstream pressure (p). Now suppose that the pressure downstream from the nozzle (p_a) is gradually reduced from its original value. The flow will be taking place in accordance with eq. (7.25) until the downstream pressure has reached a value which will be called critical. This condition corresponds to the maximum possible velocity at the throat of the nozzle, i.e. the *sonic velocity*. Further reduction in the downstream pressure will have no effect on the flow, and on the pressure at the throat (p_0) which will remain constant at its critical value. The ratio of this pressure and the upstream pressure is called the *critical pressure ratio*. Its value can be predicted from a simple equation, which may be derived as follows.

The velocity at the throat of the nozzle (v_0) is given in terms of the Mach number by eq. (7.1), from which

$$v_0 = M v_s$$

Also, from eq. (7.7), the sonic velocity v_s is given by

$$v_s = \sqrt{(k p_0 V_0)}$$

It follows that

$$v_0^2 = M^2 k p_0 V_0$$

Substituting for v_0^2 from this equation in eq. (7.24), and noting that for an isentropic flow $n = C_p/C_v = k$

$$M^2 k p_0 V_0 = \frac{2kpV}{k-1}\left[1-\left(\frac{p_0}{p}\right)^{(k-1/k)}\right]$$

from which

$$M^2 = \left(\frac{2}{k-1}\right)\left(\frac{pV}{p_0 V_0}\right)\left[1-\left(\frac{p_0}{p}\right)^{(k-1/k)}\right] \qquad (7.26)$$

Also from the law of isentropic expansion

$$\frac{V}{V_0} = \left(\frac{p_0}{p}\right)^{1/k}$$

Substituting in eq. (7.26) for

$$\frac{pV}{p_0 V_0} = \left(\frac{p}{p_0}\right)\left(\frac{p_0}{p}\right)^{1/k} = \left(\frac{p_0}{p}\right)^{(1-k)/k}$$

$$M^2 = \frac{2}{k-1}\left(\frac{p_0}{p}\right)^{(1-k)/k}\left[1-\left(\frac{p_0}{p}\right)^{(k-1)/k}\right]$$

$$M^2 = \frac{2}{k-1}\left[\left(\frac{p_0}{p}\right)^{(1-k)/k}-1\right] \qquad (7.27)$$

If the velocity in the throat of the nozzle has reached its sonic level, $M=1$, and the pressure ratio (p_0/p) becomes critical. Let (c) be the subscript denoting this condition, then from eq. (7.27)

$$\left(\frac{p_0}{p}\right)_c^{(1-k)/k} - 1 = \frac{k-1}{2}$$

from which

$$\left(\frac{p_0}{p}\right)_c^{(1-k)/k} = \frac{k+1}{2}$$

$$\left(\frac{p_0}{p}\right)_c = \left(\frac{2}{k+1}\right)^{(k/k-1)} \tag{7.28}$$

Equation (7.28) shows that the critical pressure ratio is a function of the specific heats ratio. For air and many other diatomic gases, $k=1\cdot4$ approximately.

EXAMPLE 7.5

Air leaks from a receiver into the atmosphere through a small orifice. If the pressure in the receiver is 40 psia, what is the pressure at the orifice?

Solution

Taking for air $k=1\cdot4$, the critical pressure ratio is (from eq. (7.28))

$$\left(\frac{p_0}{p}\right)_c = \left(\frac{2}{k+1}\right)^{k/(k-1)} = \left(\frac{2}{1\cdot4+1}\right)^{1\cdot4/0\cdot4}$$

$$\left(\frac{p_0}{p}\right)_c = 0\cdot528$$

Taking 14·7 psia for atmospheric pressure, the ratio $p_a/p = 14\cdot7/40 = 0\cdot37$. This is less than the critical pressure ratio, then the flow at the orifice is sonic and the pressure is critical and equal to

$$p_0 = 0\cdot528p = (0\cdot528)(40)$$
$$p_0 = 21\cdot12 \text{ psia}$$

7.8. Sonic Flow in Nozzles

Equation (7.25), developed earlier for subsonic flow, can also be used for sonic flow, provided that the pressure ratio (p_0/p) is taken at its critical value. The application of this equation to sonic flow is illustrated in the following example.

EXAMPLE 7.6

A reaction vessel is supplied with oxygen $(M_w = 32)$ from a preheater at a temperature of 30°C, through a nozzle of $\frac{1}{2}$ in. diameter throat.

The pressure upstream from the nozzle is maintained at 872 mm Hg. What is the rate of supply when the pressure in the reaction vessel is 600 mm Hg? By what percentage will the rate of supply increase if the pressure fn the reaction vessel falls to 350 mm Hg? For the oxygen, $n = 1.4$, and ior the nozzle $C_D = 0.98$.

Solution

$$p = \frac{872}{760}(14.7)(144)(32.2)$$

$$p = 78,210 \text{ poundal/ft}^2$$

$$\varrho = \frac{(32)(872)(273)}{(359)(760)(303)} = 0.0922 \text{ lb/ft}^3$$

$$A_0 = \left(\frac{1}{24}\right)^2 \frac{\pi}{4} = \frac{\pi}{2304} \text{ ft}^2$$

Let

$$k_0 = C_D A_0 \sqrt{\left(\frac{2np\varrho}{n-1}\right)}$$

$$k_0 = (0.98)\frac{\pi}{2304}\sqrt{\left(\frac{(2)(1.4)(78,210)(0.0922)}{0.4}\right)}$$

$$k_0 = 0.300$$

When the pressure in the reaction vessel is 600 mm, the flow is subcritical, and the pressure at the throat has the same value of 600 Hg as it has in the vessel, then

$$\left(\frac{p_0}{p}\right)^{2/n} = \left(\frac{600}{872}\right)^{2/1\cdot4} = 0\cdot586$$

$$\left(\frac{p_0}{p}\right)^{(n+1)/n} = \left(\frac{600}{872}\right)^{2\cdot4/1\cdot4} = 0\cdot526$$

$$\sqrt{\left\{\left(\frac{p_0}{p}\right)^{2/n} - \left(\frac{p_0}{p}\right)^{(n+1)/n}\right\}} = 0\cdot245$$

Using eq. (7.25)

$$W = (0\cdot3)(0\cdot245) = 0\cdot0735 \text{ lb/sec}$$

When the pressure in the reaction vessel falls to 350 mm, the ratio 350/872 is less then the critical pressure ratio, and the latter must be taken in eq. (7.25), thus

$$\left(\frac{p_0}{p}\right)^{2/n}_c = (0\cdot528)^{2/1\cdot4} = 0\cdot401$$

$$\left(\frac{p_0}{p}\right)^{(n+1)/n}_c = (0\cdot528)^{2\cdot4/1\cdot4} = 0\cdot335$$

$$\sqrt{(0\cdot401 - 0\cdot335)} = 0\cdot257$$

The percentage increase in flow is then

$$\frac{(0\cdot257 - 0\cdot245)100}{0\cdot245} \cong 5 \text{ per cent.}$$

7.9. Variable Flow

The process of releasing pressure in a gas container to the atmosphere (through a valve, nozzle or pipe) is an interesting example of flow under variable upstream pressure. If the initial pressure in the container is high enough, the flow will be at the critical pressure ratio, and eq. (7.25) will apply. The time required to release the pressure can be calculated from eq. (7.39), which may be derived as follows.

Equation (7.25) may be presented in the form

$$W = k_1\sqrt{(p\varrho)} \tag{7.29}$$

where the constant

$$k_1 = C_D A_o \sqrt{\left\{\frac{2n}{n-1}\left[\left(\frac{p_0}{p}\right)^{2/n} - \left(\frac{p_0}{p}\right)^{(n+1)/n}\right]\right\}} \tag{7.30}$$

Let V_c be the volume of the gas in the container, then the mass of the gas

$$m = V_c \varrho \tag{7.31}$$

But

$$\varrho = \frac{M_w}{359(T/T')(p'/p)}$$

where T' and p' are the standard temperature and pressure, respectively, while M_w is the molecular weight of the gas.

For a given gas, and isothermal flow, the last equation may be written

$$\varrho = k_2 p \tag{7.32}$$

where the new constant

$$k_2 = \left(\frac{M_w}{359}\right)\left(\frac{T'}{T}\right)\left(\frac{1}{p'}\right) \tag{7.33}$$

Substituting for ϱ from eq. (7.32) in eqs. (7.29) and (7.31)

$$W = k_1\sqrt{k_2 p} \tag{7.34}$$

$$m = k_2 V_c p \tag{7.35}$$

Let dp be the drop in pressure in the gas container when an elemental mass dm of the gas is discharged, then from eq. (7.35)

$$dm = k_2 V_c \, (dp) \tag{7.36}$$

Also let $d\theta$ be the time interval at which this change takes place, then from eq. (7.34)

$$dW = k_1\sqrt{k_2 p} \, (d\theta) \tag{7.37}$$

But since $dW = -dm$, then from the last two equations

$$k_1\sqrt{k_2 p} \, (d\theta) = -k_2 V_c \, dp$$

from which

$$d\theta = -\frac{V_c\sqrt{k_2}}{k_1}\left(\frac{dp}{p}\right) \tag{7.38}$$

Let the pressure in the container drop from p_1 to p_2 in time θ, then integrating eq. (7.38)

$$\theta = -\frac{V_c\sqrt{k_2}}{k_1}\int_1^2 \frac{dp}{p} = \frac{V_c\sqrt{k_2}}{k_1}\int_2^1 \frac{dp}{p}$$

$$\theta = \frac{V_c\sqrt{k_2}}{k_1}\log_e \frac{p_1}{p_2} \tag{7.39}$$

Finally, it will be noted that the constant k_1 includes the exponent (n) of the expansion law. This exponent is not a true constant, as it is some function of pressure, temperature, and even the size of the container. It is therefore customary to use an average value for this exponent within the pressure range involved. For air, for example, this value is often taken as 1·3 up to about 100 psig. For higher pressure, and for gases other than air, experimental values of n should be obtained.[13]

EXAMPLE 7.7

The pressure in an air receiver is released to the atmosphere through a nozzle of 0·75 in. throat diameter. The temperature of the air is 15°C and the initial pressure is 120 psia. If the volume of the receiver is 500 ft³, how long will it take to lower the pressure to 40 psia? For the nozzle, $C_D = 0·97$, and for the air, $M_w = 29$, $n = 1·3$.

Solution

The discharge through the nozzle is under critical condition throughout the pressure range, then from Example 7.5

$$\left(\frac{p_0}{p}\right)_c = 0·528$$

and

$$\left(\frac{p_0}{p}\right)_c^{2/n} = (0·528)^{2/1·3} = 0·372$$

$$\left(\frac{p_0}{p}\right)_c^{(n+1)/n} = (0·528)^{2·3/1·3} = 0·322$$

$$A_0 = \left(\frac{0·75}{12}\right)^2 \frac{\pi}{4} = \frac{\pi}{1024} \text{ ft}^2$$

From eq. (7.30)

$$k_1 = (0·97)\frac{\pi}{1024} \sqrt{\left(\frac{(2)(1·3)(0·372-0·322)}{0·3}\right)}$$

$$k_1 = 1·96 \times 10^{-3}$$

$$p' = (14·7)(144)(32·2) = 68{,}170 \text{ poundal/ft}^2$$

From eq. (7.33)

$$k_2 = \frac{(29)(273)}{(359)(288)(68170)}$$

$$k_2 = 1{\cdot}123 \times 10^{-6}$$

Using eq. (7.39)

$$\theta = \frac{(500)\sqrt{(1{\cdot}123 \times 10^{-6})}}{1{\cdot}96 \times 10^{-3}} \log_e \frac{120}{40}$$

$$\theta = 298 \text{ sec}$$

EXAMPLE 7.8

A gas leaks from a tank at such a rate that the pressure falls from 250 psia to 216 psia in 1 hour. How long will it take for the pressure to drop from 216 psia to 50 psia?

Solution

Let

$$k_0 = \frac{k_1}{V_c \sqrt{k_2}}$$

then eq. (7.39) may be written

$$k_0 \theta_1 = \log_e \frac{p_1}{p_2}$$

Taking $\theta_1 = 1$ hr

$$k_0 = \log_e \frac{250}{216}$$

$$k_0 = 0{\cdot}146$$

Now let θ be the time required in hours, then assuming that n and C_D remain constant

$$k_0 \theta = \log_e \frac{p_2}{p_3}$$

$$\theta = \frac{1}{0{\cdot}146} \log_e \frac{216}{50}$$

$$\theta = 10 \text{ hr}$$

7.10. Pneumatic Conveying

The transportation of solid materials by utilising the conveying power of gases in motion is known as *pneumatic conveying*. Air is the carrier gas commonly employed in this operation, and the main steps involved in it comprise the picking up of a solid material at one point (1), and delivering it to another distant point of the conveying system (2), where separation of the solids from the air takes place. The steps are shown diagrammatically in Fig. 7.3.

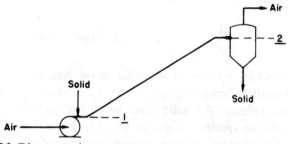

FIG. 7.3. Diagrammatic presentation of a pneumatic conveying system.

The primary requirement in the design and operation of a pneumatic conveying system is the pressure to be developed by the fan to convey the material at the desired rate. This pressure may be evaluated from a flow equation, on the assumption that the air–solid mixture is sufficiently homogeneous to obey the compressible-fluid laws. The extension of this assumption is that the total energy loss in the flow of the mixture is some function of the loss due to the flow of air alone. This leads to eq. (7.47), which is merely a refined version of similar equations suggested earlier.[14–16]

Consider, with reference to Fig. 7.3, the flow of air alone, at an average velocity v, at a section of the pneumatic line where the pressure (expressed in the absolute system of units) is p, then from the Bernoulli theorem (eq. (1.29))

$$\frac{p}{g\varrho} + \frac{v^2}{2g} + Z = \text{constant}$$

Taking $V = 1/\varrho$, and presenting this equation in a differential form

$$\frac{V\,(dp)}{g} + \frac{v\,(dv)}{g} + dZ = 0 \tag{7.40}$$

For a real fluid, let (dh_f) be the differential loss of head, then

$$\frac{V\,(dp)}{g}+\frac{v\,(dv)}{g}+dZ+dh_f = 0 \tag{7.41}$$

It will be noted that each term of this equation is based on one pound of the air used in the conveying line.

Now, let r be the number of pounds of solid material conveyed by each pound of air, then treating the gas–solid mixture as a homogeneous fluid, and assuming that the solid particles have the same velocity as the air, the following flow equation may be written for the r pounds of solid particles.

$$rV_s\,\frac{dp}{g}+r\,\frac{v\,(dv)}{g}+r\,(dZ)+(dh_f)_s = 0 \tag{7.42}$$

where the subscript s refers to the solid phase, and the term $(dh_f)_s$ represents the contribution of this phase to the total loss of energy.

The specific volume of a solid material (V_s) is negligibly small when compared with the specific volume of air (V), thus the first term of the equation may be ignored. Adding the remaining terms of this equation to eq. (7.41)

$$V\,\frac{dp}{g}+(1+r)\,\frac{v\,(dv)}{g}+(1+r)\,dZ+dh_f+(dh_f)_s = 0 \tag{7.43}$$

Let the total loss of energy be given by

$$dh_f+(dh_f)_s = \alpha\,(dh_f) \tag{7.44}$$

where α is a factor denoting the number of times the total energy loss is greater than that experienced in the flow of air alone.

Making use of the Darcy equation, the energy loss in the flow of air at a velocity v is

$$h_f = 4f\,\frac{L}{D}\,\frac{v^2}{2g}$$

and

$$dh_f = 4f\,\frac{L}{D}\,\frac{v\,(dv)}{g}$$

Substituting for (dh_f) from this equation in eq.(7.44), and from the resulting equation in eq. (7.43)

$$V\left(\frac{dp}{g}\right)+(1+r)\,\frac{v(dv)}{g}+(1+r)\,dZ+\alpha\left(\frac{4fL}{gD}\right)v\,(dv) = 0 \tag{7.45}$$

It has been shown earlier that

$$v\,(dv) = G^2 V\,(dV)\tag{7.15}$$

Also, from the ideal-gas law

$$V = \frac{RT}{p}$$

and

$$dV = -\frac{RT}{p^2}\,(dp)$$

then

$$V(dV) = -\frac{(RT)^2}{p^3}\,(dp)$$

and

$$v(dv) = -\frac{(GRT)^2}{p^3}\,(dp)$$

Substituting from the respective equations for V and $v(dv)$ in eq. (7.45)

$$\frac{RT}{g}\left(\frac{dp}{p}\right)-(1+r)(GRT)^2\frac{dp}{p^3}+(1+r)dZ-\alpha\left(\frac{4fL}{gD}\right)(GRT)^2\frac{dp}{p^3}=0$$

$$\frac{RT}{g}\left(\frac{dp}{p}\right)-(GRT)^2\left[(1+r)+\alpha\left(\frac{4fL}{gD}\right)\right]\frac{dp}{p^3}+(1+r)dZ=0$$

$$\tag{7.46}$$

Integrating between the limits 1 and 2 (see Fig. 7.3)

$$\frac{RT}{g}\log_e\frac{p_1}{p_2}+\frac{(GRT)^2}{2}\left[(1+r)+\alpha\left(\frac{4fL}{gD}\right)\right]\left[\frac{1}{p_1^2}-\frac{1}{p_2^2}\right]+(1+r)(Z_1-Z_2)=0$$

$$\tag{7.47}$$

A great deal of research has been conducted[46] in an attempt to correlate experimental data for the factor α. The resulting equations have not, however, proved to be simple enough for general use, and not necessarily reproducible for systems other than those used in the tests. An arbitrary value will therefore be assigned to this factor in the following example, in order to illustrate the application of eq. (7.47) in an industrial problem.

EXAMPLE 7.9

A granular material is to be conveyed pneumatically in a line of 6 in. diameter and 420 ft long, at a rate of 20,000 lb/hr. The discharge end of the line will be open to the atmosphere, 80 ft above the feed point, and it is intended to maintain a 10 to 1 solid–air mass ratio. If, under the operating conditions, the factor α has a value of 20·3, what pressure will be required at the feed point? A Darcy friction factor of 0·006, and 18°C for the temperature of the air may be assumed.

Solution

For the conveying of 20,000 lb/hr of solids, at $r=10$, the rate of flow of air is 2000 lb/hr, and

$$G = \frac{2000}{(3600)\,(0·5)^2\pi/4}$$

$$G = 2·83 \text{ lb/(ft}^2)\text{ (sec)}$$

Using the absolute system of units, and taking 29 for the molecular weight of air

$$RT = pV = (14·7)\,(144)\,(32·2)\left(\frac{359}{29}\right)\left(\frac{273+18}{273}\right)$$

$$RT = 8·995\times10^5\,\frac{\text{(poundal) (ft)}}{\text{lb}}$$

$$\frac{RT}{g} = \frac{8·995\times10^5}{32·2} = 2·794\times10^4$$

$$(GRT)^2 = [(2·83)\,(8·995\times10^5)]^2 = 6·48\times10^{12}$$

$$p_2 = (14·7)\,(144)\,(32·2)$$

$$p_2 = 6·817\times10^4\,\frac{\text{poundal}}{\text{ft}^2}$$

$$\frac{4fL}{gD} = \frac{(4)\,(0·006)\,(420)}{(32·2)\,(0·5)} = 0·6244$$

Substituting in eq. (7.47)

$$2·794\times10^4 \log_e \frac{p_1}{6·817\times10^4} + \frac{6·48\times10^{12}}{2}\,[(1+10)+(20·3)\,(0·6244)]$$

$$\times\left[\frac{1}{p_1^2} - \frac{1}{6·817\times10^4}\right] + (1+10)\,(-80) = 0$$

from which

$$p_1 = 8·930\times10^4\,\frac{\text{poundal}}{\text{ft}^2}$$

This is equivalent to 19·7 psia.

Problems

7.1. Air ($M_w = 29$, $k = 1\cdot4$) flows through a Venturi tube of 1 in. diameter throat at a rate of 2 cfs, measured at the throat, where the pressure and temperature are 14·7 psia, and 17°C, respectively. What is the velocity at the throat? At what Mach number does the flow take place?

(*Ans.*: 367 fps, 0·32)

7.2. Hydrogen ($M_w = 2$, $k = 1\cdot4$) flows through a nozzle of 0·5-in. diameter throat at a Mach number of 0·1. If the temperature and pressure at the throat are 25°C, and 29·4 psia, respectively, what is the flow?

(*Ans.*: 0·361 lb/min)

7.3. A natural gas ($M_w = 16$) is to flow through a horizontal pipe of 24 in. diameter at a mass velocity of 12 lb/ft²×sec, against an outlet pressure of 15 psia. If the Darcy friction factor is 0·01, and the temperature of the gas is 17°C, what will be the drop in pressure in a length of 6 miles (31,680 ft)?

(*Ans.*: 69·34 psia)

7.4. A natural gas ($M_w = 16$) is at a steady flow to a storage tank through a horizontal pipe of 6 in. diameter and 20,000 ft long. The pressure in the tank is maintained at 16·7 psia. If the pressure at the inlet to the pipe is 74·7 psia, what is the flow? The gas is at a uniform temperature of 17°C, and the Darcy friction factor is 0·01.

(*Ans.*: 78·0 lb/min)

7.5. A horizontal pipe, of $\frac{1}{2}$ in. diameter and 55 ft long, carries superheated steam ($M_w = 18$) from the steam-main to a point where the pressure is 14·7 psia. If the steam flows at a constant temperature of 270°F, and the pressure in the main is 28·7 psia, what is the flow? The Darcy friction factor is 0·008.

(*Ans.*: 3·43 lb/min)

7.6. A gaseous fuel ($M_w = 16$) is at a steady flow through a nozzle of 1 in. diameter throat to a furnace where the pressure is 14·7 psia. The temperature and pressure of the gas at the entrance to the nozzle are 17°C, and 24 psia, respectively. If the expansion factor is $n = 1\cdot4$, and for the nozzle $C_D = 0\cdot95$, what is the flow? (*Ans.*: 17·82 lb/min)

7.7. A nozzle of 1 in. diameter throat is fitted into an air receiver where the temperature is 17°C. If the pressure in the receiver is 120 psia, what is the flow into the atmosphere? For the air $M_w = 29$, $n = 1\cdot4$, and the critical pressure ratio is 0·528. For the nozzle $C_D = 0\cdot96$.

(*Ans.*: 2·1 lb/sec)

7.8. Air ($M_w = 29$) discharges from a receiver of 400 ft³ volume into the atmosphere through a valve equivalent to a nozzle of $\frac{1}{4}$ in. diameter throat, and with a coefficient of discharge of 0·95. If the temperature of the air is 15°C, and the expansion obeys the law $pV^{1\cdot3}$ = constant, how long will it take to release the pressure from 95 to 45 psia?

(*Ans.*: 660 sec)

7.9. A leak in a gas container results in a drop in pressure from 92 psia to 88 psia in $1\frac{1}{2}$ hr. How many more hours will be required for the pressure to fall to 50 psia? The coefficient of discharge, and the expansion law index may be assumed constant.

(*Ans.*: 25·12 hr)

7.10. A solid material is conveyed pneumatically in a line of 6 in. diameter, and 300 ft long, the discharge end being open to the atmosphere 40 ft above the feed point. In one run, the following information was recorded.

$$\begin{aligned}
\text{Atmospheric pressure} &= 14\cdot7 \text{ psia} \\
\text{Feed point pressure} &= 18\cdot5 \text{ psia} \\
\text{Solid–air mass ratio} &= 4 \\
\text{Air mass velocity} &= 5\cdot8 \text{ lb/(ft}^2)(\text{sec}) \\
\text{Air temperature} &= 18°\text{C}
\end{aligned}$$

Assuming the Darcy friction factor $f = 0\cdot005$, what is the value of the factor α of eq. (7.47) under these conditions?   (*Ans.*: $2\cdot1$)

CHAPTER 8

FLOW THROUGH EQUIPMENT

FLOW past a solid boundary other than the inside wall of a pipe, or channel bed, is of considerable interest in industrial practice. Heat exchangers, chimneys, filters and packed columns are among the equipment where this kind of flow occurs.

8.1. Pressure Drop outside Tubes

The determination of the pressure drop in flow outside tubes is an essential requirement in the design of heat transfer equipment. Out of the many kinds of equipment used in industry for the purpose of heating, cooling or exchanging heat between two fluid streams, the so-called shell-and-tube type is most common. Its essential features are outlined in the following discussion, and reference is made to the relevant literature[17, 18] for more detailed description.

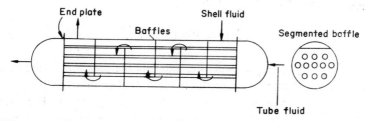

FIG. 8.1. Flow outside a bank of tubes in a single pass, shell-and-tube-type heat exchanger.

The standard shell-and-tube-type heat exchanger consists of a cylindrical shell, normally cut from a piece of pipe of larger size, and a number of rows of tubes fitted in the end plates, as shown in Fig. 8.1. The shell side

233

fluid is made to flow up and down across these tubes and between baffles, which are plates laid across the shell, and held securely by means of special tie-rods. The baffles may be shaped in a variety of ways but the segmental type, with a cutting made approximately 25 per cent of the plate area, is most common.

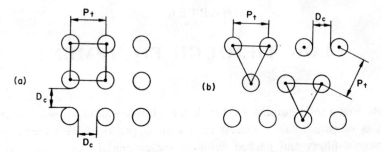

FIG. 8.2. Tube layout, (a) on a square pitch (in-line arrangement), (b) on a triangle pitch (staggered arrangement).

The tubes may be laid on either a square or triangular pitch, as shown in Fig. 8.2. For the benefit of the heat transfer and pressure drop characteristics, their spacing should comply with some standards. These require that the pitch, which is the transverse distance between the centres of adjacent tubes, has a range of between $1 \cdot 25 \, D_0$ and $1 \cdot 5 \, D_0$, where D_0 is the outside diameter of the tubes.

The determination of the drop in pressure outside the tubes is commonly approached by adapting one of the equations, such as the Darcy equation, used for flow in pipes. Let N be the number of baffles, and N_r be the number of rows of tubes in a single shell, then

$$(\Delta p)_s = 4fN_r(N+1)\frac{v^2 \varrho}{2g_c} \qquad (8.1)$$

where $(\Delta p)_s$ is the drop in pressure, and v is the maximum fluid velocity, based on the minimum area available for flow. The new friction factor (f) is not as easy to determine as it is in flow through pipes. The common practice is to evaluate it with the aid of some experimentally obtained diagrams,[17] which are usually plots of this factor and the Reynolds number based on a strictly specified linear dimension.

For banks of tubes in a staggered arrangement only, the friction factor is simply related to the Reynolds number by the equation

$$f = 0 \cdot 75 Re^{-0 \cdot 2} \qquad (8.2)$$

when

$$Re = \frac{D_c v \varrho}{\mu} \tag{8.3}$$

where D_c is the transverse clearance between the tubes, and v is the maximum velocity through the minimum flow area.

Unless the tubing layout is known, the minimum flow area is difficult to determine. This difficulty is often obviated by calculating this area at the plane passing through the centre of the shell, despite the fact that in most layouts there is no row of tubes at this plane. The minimum flow area is therefore based on a hypothetical row of tubes at the shell diameter.

Let D_s be the inside diameter of the shell, P_t be the transverse pitch, and B be the baffle spacing, (see Fig. 8.3), then

$$A_s = D_s B \frac{D_c}{P_t} \tag{8.4}$$

where A_s is the required area for the evaluation of the maximum velocity across the tubes.

The drop in pressure across a segmental type baffle may be calculated approximately[17] on the assumption that its opening represents an orifice with a coefficient of discharge of 0·7, as defined by eq. (8.5). This equation may be derived as follows.

Let v be an average velocity through the opening, then from an orifice equation

$$v = \sqrt{(2gH)} \tag{1.40}$$

Substituting in this equation for

$$H = \frac{\Delta p}{\gamma} = \frac{\Delta p}{\varrho(g/g_c)}$$

and

$$v = \frac{G_b}{\varrho}$$

$$\frac{G_b}{\varrho} = \sqrt{\left(2g \frac{(\Delta p)g_c}{\varrho g}\right)}$$

from which

$$G_b = \sqrt{(2g_c(\Delta p)\varrho)}$$

or

$$\Delta p = \frac{G_b^2}{2g_c\varrho}$$

where G_b is the mass velocity across the baffle opening.

Let C_D be the coefficient of discharge, then for N baffles, the drop in pressure

$$(\Delta p)_b = \frac{NG_b^2}{2C_D g_c \varrho} \tag{8.5}$$

EXAMPLE 8.1

A liquid is cooled outside the tubes of a single-pass, shell-and-tube exchanger, at a rate of 180,000 lb/hr. The shell has a diameter of 25 in., and the tubes are 8 ft long and of 1 in. outside diameter. They are arranged in twelve staggered vertical rows at $1\frac{1}{4}$ in pitch. The baffles are of segmental type with 25 per cent area cutting, and spaced 12 in. apart. If the liquid has an average density and viscosity of 60 lb/ft³, and 2·4 (lb/ft) (hr) respectively, what is the drop in pressure?

Solution

The transverse clearance is

$$D_c = P_t - D_0 = 1 \cdot 25 - 1 \cdot 00 = 0 \cdot 25 \text{ in.}$$

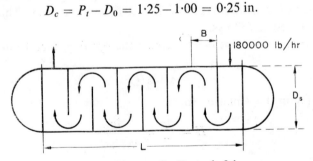

FIG. 8.3. Diagram for Example 8.1.

Using eq. (8.4), the flow area at the shell diameter is

$$A_s = D_s B \frac{D_c}{P_t} = \left(\frac{25}{12}\right) \left(\frac{12}{12}\right) \left(\frac{0 \cdot 25}{1 \cdot 25}\right)$$

$$A_s = \frac{5}{12} \text{ ft}^2$$

The maximum mass velocity

$$G_s = \frac{180,000}{5/12} = 432,000 \frac{\text{lb}}{(\text{ft}^2) (\text{hr})}$$

Since $G_s = v\varrho$, then from eq. (8.3)

$$Re = \frac{D_c G_s}{\mu} = \frac{(0 \cdot 25/12)(432,000)}{2 \cdot 4}$$

$$Re = 3750$$

Using eq. (8.2)

$$f = 0 \cdot 75 Re^{-0 \cdot 2} = (0 \cdot 75)(3750)^{-0 \cdot 2}$$
$$f = 0 \cdot 145$$

The velocity at shell diameter

$$v = \frac{G_s}{3600\varrho} = \frac{432,000}{3600\varrho}$$

$$v = \frac{120}{\varrho} \text{ fps}$$

Substituting for v from this equation in eq. (8.1), and for $\varrho = 60$ lb/ft³

$$(\varDelta p)_s = 4fN_r(N+1)\frac{(120)^2}{2g_c\varrho}$$

$$(\varDelta p)_s = 4(0 \cdot 145)(12)(7+1)\frac{(120)^2}{(2)(32 \cdot 2)(60)}$$

$$(\varDelta p)_s = 208 \text{ psf}$$

The flow area across one baffle with 25 per cent cutting

$$A_b = (0 \cdot 25)\frac{\pi D_s^2}{4} = (0 \cdot 25)\left(\frac{25}{12}\right)^2\left(\frac{\pi}{4}\right)$$

$$A_b = 0 \cdot 851 \text{ ft}^2$$

$$G_b = \frac{180,000}{0 \cdot 851} = 211,300 \text{ lb/(ft}^2)\text{ (hr)}$$

Substituting in eq. (8.5), and taking $C_D = 0 \cdot 7$, $N = 7$ baffles, and $\varrho = 60$ lb/ft³

$$(\varDelta p)_b = \frac{(7)(211,300)^2}{(2)(32 \cdot 2)(3600^2)(0 \cdot 7)(60)}$$

$$(\varDelta p)_b = 8 \cdot 93, \text{ say } 9 \text{ psf}$$

Total pressure drop

$$\varDelta p = (\varDelta p)_s + (\varDelta p)_b = 208 + 9$$
$$\varDelta p = 217 \text{ psf}$$

8.2. The Chimney Effect

With the exception of the domestic type, the primary objective of a chimney is to disperse industrial polution to an acceptable concentration at ground level. From this standpoint, chimneys as high as 800 ft are known to have been built, but generally, the heights of chimneys are limited by such factors as cost, nuisance to aircraft, and objections on aesthetic grounds.

The secondary objective of an industrial chimney is to create a draught, and it is this function which attracts attention in the subject of fluid mechanics. The magnitude of the draught is simply related to the chimney height and to the average density of the gases it conveys. The interrelationship may be derived as follows.

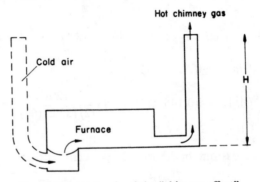

FIG. 8.4. An example of the "chimney effect".

Consider, in Fig. 8.4, an industrial furnace discharging hot gases into the atmosphere through a chimney of height H. Let ϱ_h be the average density of the hot gases in the chimney, and ϱ be the density of the ambient air, then by a balance of forces on a unit cross-section basis

$$(\Delta p)g_c = H(\varrho - \varrho_h)\,g \tag{8.6}$$

where (Δp) is the differential presure between the inlet to the furnace and the outlet from the chimney, i.e. the "draught" produced by the so called "chimney effect". This effect may be visualised as being equivalent to the difference in weight between the columns of ambient air (bounded in Fig. 8.4 by broken lines) and the chimney gases, both columns being of the same height and cross-sectional area.

If the density of the ambient air is assumed constant, then for a given height of the chimney the draught is simply related to the average density of the hot chimney gases, by eq. (8.6). From the Bernoulli theorem, we should also expect some effect of the velocity head inside the chimney on the draught. If the average velocity of the chimney gas is denoted by the symbol v, the combined effect may be presented in the form of the following proportionality.

$$\Delta p \propto (\varrho_h)\left(\frac{v^2}{2g_c}\right)$$

or

$$\frac{\Delta p}{\varrho_h} = k\left(\frac{v^2}{2g_c}\right) \qquad (8.7)$$

where k is a proportionality constant.

Substituting in the last equation from eq. (8.6) for

$$\Delta p = H(\varrho - \varrho_h)\frac{g}{g_c}$$

and cancelling in the resulting equation the g_c terms

$$\frac{H(\varrho - \varrho_h)g}{\varrho_h} = k\left(\frac{v^2}{2}\right)$$

Let G be the mass-velocity of the chimney gases, then substituting in the last equation for

$$v = \frac{G}{\varrho_h}$$

$$\frac{H(\varrho - \varrho_h)g}{\varrho_h} = k\left(\frac{G^2}{2\varrho_h^2}\right)$$

Simplifying

$$H(\varrho - \varrho_h)g = k\left(\frac{G^2}{2\varrho_h}\right)$$

from which

$$G^2 = \frac{2gH(\varrho - \varrho_h)\varrho_h}{k}$$

and

$$G = \sqrt{\left(\frac{2gH}{k}\right)} \sqrt{(\varrho - \varrho_h)\varrho_h} \qquad (8.8)$$

Equation (8.8) provides a convenient basis for the determination of the most economical operational conditions for chimneys, i.e. the conditions

at which the mass-velocity of the hot chimney gas is a maximum, i.e.
when $dG/d\varrho_h = 0$, or

$$\frac{d[(\varrho-\varrho_h)\varrho_h]}{d\varrho_h} = \varrho-2\varrho_h = 0$$

$$\text{from which } \varrho_h = \frac{\varrho}{2} \tag{8.9}$$

Equation (8.9) indicates that chimneys operate most economically when
the density of the gases they convey is half the density of the ambient
air. Since the density of gases is inversely proportional to their absolute
temperatures, then—taking 70°F for the atmospheric air—the optimum
temperature of a chimney gas is found to be $2(460+70) = 1060°R$, or
600°F, approximately. This temperature is, however, rarely aimed at in
the operation of chimneys, since the usual requirement is that heat is
extracted from the hot flue gases as efficiently as possible before the gases
enter the chimney.

The determination of the average density of chimney gases should take
into account the cooling effect to which they are subjected on their way
to the atmosphere. This effect is due to heat losses by conduction through
the walls of the chimney, and also due to the leakage of atmospheric air.[19]
The latter also increases the volume of the chimney gases, and for this
reason the temperature drop is difficult to determine. It is normally
aproximated by ignoring the small change in the mass velocity, and assuming a 0·5°F drop for brick structures, or a 1°F drop for steel chimneys,
per foot of height.

The determination of the cross-sectional area of chimneys is based on
economic grounds, and this problem may be approached in the same
way as outlined in Chapter 3 for pipes. More commonly, however, the
economic cross-section of a chimney is assessed approximately from
experimental data, and Table 8.1[19] may serve as a guide in selecting
the economic velocity from which the cross-section can be calculated.

TABLE 8.1

Chimney capacity (cfs)	Economic velocity (fps)
10	10
50	15
150	20

EXAMPLE 8.2

A steel chimney is to be erected to convey flue gases through a fire-tube boiler, containing 120 tubes of 2·9 in. diameter and 10 ft long, arranged in a single pass. The flue gases ($M_w = 30$) will enter the boiler at 1800°F and leave it for the chimney at 620°F at a flow rate of 325 lb/min. If the Darcy friction factor for the tubes is 0·01, and 20 per cent is allowed for losses other than tube friction, what should be the height and cross-section of the chimney?

Solution

Let t_1 be the mean temperature of the flue gases passing through the boiler tubes, then

$$t_1 = \frac{1800 + 620}{2} = 1210°F$$

The mean density of the gases ($M_w = 30$) at this temperature is

$$\varrho_1 = \frac{(30)(492)}{(359)(460 + 1210)}$$

$$\varrho_1 = 0·0248 \text{ lb/ft}^3$$

and the volumetric rate of flow

$$Q_1 = \frac{325}{(60)(0·0248)} \text{ ft}^3/\text{sec}$$

For 120 tubes, the flow area is

$$A_1 = 120 \left(\frac{2·9}{12}\right)^2 \left(\frac{\pi}{4}\right) \text{ ft}^2$$

The mean velocity through the tubes of the boiler is then

$$v_1 = \frac{Q_1}{A_1} = \frac{(325)(4)}{(60)(0·0248)(120)(2·9/12)^2(\pi)}$$

$$v_1 = 43·4 \text{ fps}$$

Using the Darcye quation, with a 20 per cent allowance made for losses other than tube friction

$$(\Delta p) = 4f \left[\frac{1 \cdot 2L}{D} \right] \left[\frac{v_1^2 \varrho_1}{2g_c} \right]$$

$$= 4(0 \cdot 01) \left[\frac{1 \cdot 2 \times 10}{2 \cdot 9/12} \right] \left[\frac{(43 \cdot 4)^2 (0 \cdot 0248)}{64 \cdot 4} \right]$$

$$\Delta p = 4 \cdot 186 \text{ psf}$$

Let H be the required height of the chimney, then taking a $1°F$ temperature drop per foot of height, the mean temperature of the chimney gas is

$$t_h = \frac{620 + (620 - 1 \cdot 0H)}{2}$$

$$t_h = 620 - \frac{H}{2}$$

and the mean density of the chimney gas

$$\varrho_h = \frac{(30)(492)}{(359)(460 + 620 - H/2)}$$

$$\varrho_h = \frac{82 \cdot 24}{2160 - H}$$

Taking for the density of the atmospheric air $\varrho = 0 \cdot 078 \text{ lb/ft}^3$

$$\varrho - \varrho_h = 0 \cdot 078 - \frac{82 \cdot 24}{2160 - H}$$

$$\varrho - \varrho_h = \frac{86 \cdot 3 - 0 \cdot 078H}{2160 - H}$$

Substituting in eq. (8.6), for (Δp) and $(\varrho - \varrho_h)$ from the respective equations, and noting that the terms g and g_c have the same numerical values

$$\frac{H(86 \cdot 3 - 0 \cdot 078H)}{2160 - H} = 4 \cdot 186$$

from which

$$H = 125 \text{ ft}$$

(It will be noted that the other root of the equation would lead to an absurd solution.)

Substituting the obtained value for the height of the chimney in the respective equations gives

$$t_h = 620 - \frac{125}{2} = 557 \cdot 5°F$$

$$\varrho_h = \frac{(30)\,(492)}{(359)\,(460 + 557 \cdot 5)}$$

$$\varrho_h = 0 \cdot 0404 \ \text{lb/ft}^3$$

The mean volumetric rate of flow of the chimney gas is therefore

$$Q = \frac{325}{(60)\,(0 \cdot 0404)}$$

$$Q = 134 \ \text{ft}^3/\text{sec}$$

For this capacity, the economic velocity is nearly 20 fps (see Table 8.1). Taking this value, the economic cross-sectional area of the chimney is approximately $134/20 = 6 \cdot 7 \ \text{ft}^2$.

8.3. Pressure Drop through Beds of Solid Particles

The first scientific approach to the study of flow through beds of solid particles is attributed to Darcy. Apart from his major contribution to hydraulics in the field of fluid friction in pipes, Darcy is also noted for his work on the flow of water through beds of sand. The result of his investigations was published in 1856 in a treatise[20] which is generally regarded as the first recorded work on the subject. The leading conclusion drawn by Darcy from the results of his investigations was that the drop in pressure through a bed of solid particles was proportional to the first power of the rate of flow rather than, as it was then popularly supposed,[1] to its square root.

A number of attempts have been made[21] since the time of Darcy to express the drop in pressure in terms of the rate of flow and physical properties of the fluid–solid systems. Among these was a correlation of experimental data, presented in 1931 by Chilton and Colburn,[22] which attracted a great deal of attention. Based on the Darcy equation for flow in pipes, the correlation assumes the form

$$\Delta p = \frac{4f'h\varrho v^2 F_w F_p}{2g_c D_p} \tag{8.10}$$

where Δp = pressure drop,
    v = superficial velocity, based on empty cross-section,
    D_p = nominal size of a packing particle,
    h = depth of packing,
    F_w = wall-effect factor,
    F_p = correction factor for hollow particles.

Introducing the concept of mass-velocity, $\varrho v^2 = G^2/\varrho$, eq. (8.10) becomes

$$\Delta p = \frac{4f'hG^2F_wF_p}{2g_c\varrho D_p} \tag{8.11}$$

The friction factor f', which appears in the above equations, may be obtained from a diagram (Fig. 8.5), which is a plot of this factor against a modified Reynolds number defined by

$$N_{Re} = \frac{D_p v\varrho}{\mu} = \frac{D_p G}{\mu} \tag{8.12}$$

Figure 8.5 is the outcome of an excellent survey of the data available on geometrically regular shapes as well as on irregularly shaped particles.

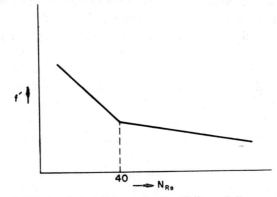

FIG. 8.5. Friction factor diagram for the Chilton–Colburn equation.

The original diagram shown a curve drawn through widely scattered points, particularly in the region of high Reynolds numbers. For the sake of simplicity, however, the curve has been approximated by two straight lines intersecting at a point corresponding to $N_{Re} = 40$, which has been accepted as the transition point between the viscous and turbulent flow regions. The two lines are described approximately by the following equations.

For the streamline region ($N_{Re} < 40$)

$$f' = \frac{850}{N_{Re}} \qquad (8.13)$$

and for the turbulent region ($N_{Re} > 40$)

$$f' = \frac{38}{N_{Re}^{0.15}} \qquad (8.14)$$

The wall-effect factor (F_w), which appears in eq. (8.10), has been determined experimentally by Furnas,[23] for solid particles only, and presented as a plot against the ratio of particle–column diameter. This factor may be taken as unity for hollow shapes, provided that another correction factor (F_p) is used instead. From the correlation of data presented by White,[24] this factor may be evaluated in terms of the nominal particle size (D_p) from the equation

$$F_p = \frac{C}{D_p^{0.5}}$$

The constant C in this equation has a value of 0·24 for Raschig or Lessing rings, and 0·13 for Berl saddles.

EXAMPLE 8.3

A gas ($\varrho = 0.070$ lb/ft^3, $\mu = 0.042$ lb/(ft)(hr)) is at a steady flow of 882 lb/(ft^2)(hr) through a bed formed by cubes of 2 in. side. If the wall effect factor is 0·9, what is the drop in pressure for a depth of 3 ft?

Solution

Using eq. (8.12)

$$N_{Re} = \frac{(2/12)(882)}{0.042} = 3500$$

Since the Reynolds number exceeds 40, eq. (8.14) applies, hence

$$f' = \frac{38}{N_{Re}^{0.15}} = \frac{38}{3500^{0.15}}$$

$$f' = 11 \cdot 2$$

For solid particles, the correction factor $F_p = 1$, and using eq. (8.11)

$$\Delta p = \frac{(4)\,(11\cdot2)\,(3)\,(882/3600)^2\,(0\cdot9)}{(2)\,(32\cdot2)\,(0\cdot07)\,(2/12)}$$

$$\Delta p = 9\cdot66 \text{ psf}$$

or

$$\frac{(9\cdot66)\,(12)}{62\cdot4} = 1\cdot86 \text{ i.w.g.}$$

8.4. Kozeny's Hydraulic Model

The flow through a bed of solid particles takes place in the irregularly shaped channels provided by the voids of the bed. It seems therefore logical that any equation relating to this kind of flow should refer to these channels rather than to the size of the bed forming particles, as suggested by eq. (8.11). The equation which takes this into account is due to Carman and Kozeny. It is the outcome of a theory based on a hydraulic model devised by Kozeny.[25] The Kozeny hydraulic model assumes that the channels are formed by capillary tubes running parallel to the direction of flow, as shown in Fig. 8.6.

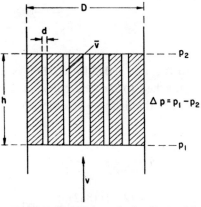

Fig. 8.6. The Kozeny hydraulic model.

Consider, in Fig. 8.6, the flow of a fluid through a bed of solid particles contained in a cylindrical vessel of diameter D. Assume that the flow takes place through cylindrical channels of diameter d and length h, which is also the depth of the bed. Let n be the total number of these channels, then from the definition of the voidage fraction

$$\varepsilon = \frac{\text{volume of voids}}{\text{volume of bed}} = \frac{n \, d^2(\pi/4)\,(h)}{D^2(\pi/4)\,(h)}$$

$$\varepsilon = n\left(\frac{d}{D}\right)^2 \tag{8.15}$$

Now, let $\bar{v}$ be the average velocity through the parallel channels, then by the law of continuity

$$vD^2\,\frac{\pi}{4} = n\bar{v}d^2\,\frac{\pi}{4}$$

from which

$$n = \frac{v}{\bar{v}}\left(\frac{D}{d}\right)^2$$

Substituting for n from this equation in eq. (8.15), and simplifying

$$\bar{v} = \frac{v}{\varepsilon} \tag{8.16}$$

Similarly, the hydraulic mean radius may be expressed in terms of the voidage fraction as follows.

Let V_p be the volume of a single particle, and A_p be its surface area, then in a uniformly sized bed of N particles

$$NV_p = \text{total volume of particles,}$$
$$NA_p = \text{total surface area of particles,}$$

and

$$NV_p\left(\frac{\varepsilon}{1-\varepsilon}\right) = \text{total volume of voids.}$$

The hydraulic mean radius is generally defined as the ratio of the cross-sectional area and the wetted perimeter. For a bed of solid particles, this parameter is conveniently defined as

$$R_H = \frac{\text{Total volume of voids}}{\text{Total surface area of particles}}$$

It follows that

$$R_H = \frac{NV_p\big(\varepsilon/(1-\varepsilon)\big)}{NA_p} = \frac{V_p\varepsilon}{A_p(1-\varepsilon)}$$

Let S_p be the specific area of a single particle, then

$$S_p = \frac{A_p}{V_p} \qquad (8.17)$$

and

$$R_H = \frac{\varepsilon}{S_p(1-\varepsilon)} \qquad (8.18)$$

The Reynolds number may be defined in terms of R_H, as the characteristic linear dimension, by

$$Re = \frac{R_H \bar{v} \varrho}{\mu} \qquad (8.19)$$

Substituting in this equation for $\bar{v}$ and R_H from the respective equations (8.16) and (8.18)

$$Re = \frac{[\varepsilon/S_p(1-\varepsilon)]\,(v/\varepsilon)\,(\varrho)}{\mu}$$

Simplifying

$$Re = \frac{v\varrho}{S_p(1-\varepsilon)\mu} \qquad (8.20)$$

or, taking $G = v\varrho$

$$Re = \frac{G}{S_p(1-\bar{\varepsilon})\mu} \qquad (8.21)$$

EXAMPLE 8.4

A reactor contains particles shaped into cylinders of 0·5 in. diameter and 0·8 in. height. If a gas ($\varrho = 0\cdot08$ lb/ft³, $\mu = 0\cdot042$ lb/(ft) (hr)) is at a steady flow through the bed ($\varepsilon = 0\cdot4$) at a superficial velocity of 2 fps, what is the corresponding Reynolds number?

Solution

$$A_p = (2)\,(0\cdot5)^2\,(\pi/4)+(0\cdot5)\,(\pi)\,(0\cdot8)$$

$$A_p = \frac{21\pi}{40} \quad \text{in}^2$$

$$V_p = (0\cdot5)^2\,(\pi/4)\,(0\cdot8)$$

$$V_p = \frac{\pi}{20} \quad \text{in}^3$$

Using eq. (8.17)

$$S_p = \frac{A_p}{V_p} = \frac{21\pi/49}{\pi/20} = 10{\cdot}5 \text{ in}^2/\text{in}^3$$

$$S_p = 126 \text{ ft}^2/\text{ft}^3$$

Substituting the data in eq. (8.20)

$$Re = \frac{(3600)\,(2)\,(0{\cdot}08)}{(126)\,(1-0{\cdot}4)\,(0{\cdot}042)}$$

$$Re = 181$$

8.5. The Carman–Kozeny Equation

The drawbacks inherent in eq. (8.10), which is simply a form of the Darcy equation for flow in pipes, can be removed by adapting it to the theory advanced by Kozeny, as outlined in the preceding paragraph.

Ignoring the factors F_p and F_w in eq. (8,4), and replacing the terms v and D_p in this equation by their equivalents $\bar{v}$ and R_H, leads to

$$\Delta p = \frac{4f''h\varrho\bar{v}^2}{2g_c R_H}$$

where f'' is a new friction factor.

Again, substituting in this equation for $\bar{v}$ and R_H from the respective equations (8.16) and (8.18), and letting $f_c = 2f''$

$$\Delta p = \frac{f_c h\varrho(v/\varepsilon)^2}{g_c[\varepsilon/S_p(1-\varepsilon)]}$$

Rearranging the terms of this equation

$$\Delta p = f_c \left(\frac{v^2}{g_c}\right)\left(\frac{1-\varepsilon}{\varepsilon^3}\right)(h\varrho S_p) \tag{8.22}$$

This equation, as well as eq. (8.23), is known as the Carman–Kozeny equation. The latter is obtained by substituting in eq. (8.22) for

$$v = \frac{G}{\varrho}$$

so that

$$\Delta p = f_c \left(\frac{G^2}{g_c}\right)\left(\frac{1-\varepsilon}{\varepsilon^3}\right)\left(\frac{hS_p}{\varrho}\right) \tag{8.23}$$

From the available data, Carman made a logarithmic plot of the factor f_c, now called the Carman friction factor, against the Reynolds number, as defined by eq. (8.21). The resulting line indicated a linear relationship up to $Re = 4$, approximately. From the slope of the straight portion of the curve, Carman deduced the following relationship between the two variables.

$$f_c = \frac{5}{Re} \tag{8.24}$$

By analogy to flow in pipes, the range up to $Re = 4$, to which eq. (8.24) applies, is called laminar or viscous. For the same reason, the range above $Re = 4$ is called turbulent. Within this range, the line on the Carman plot curves moderately, and can be approximately described by the equation

$$f_c = \frac{5}{Re} + \frac{0\cdot4}{Re^{0\cdot1}} \tag{8.25}$$

Further experimental evidence has confirmed the validity of eq. (8.25) for a wide range of particle sizes, with the exception of hollow shapes, for which the following equation has been suggested[26]:

$$f_c = \frac{5}{Re} + \frac{1}{Re^{0\cdot1}} \tag{8.26}$$

For spherical particles, the specific surface is given (from eq. (8.13), taking $A_p = \pi D_p^2$, and $V_p = \pi D_p^3/6$) by

$$S_p = \frac{6}{D_p} \tag{8.27}$$

where D_p is the particle diameter.

Substituting from this equation for S_p in eq. (8.20)

$$Re = \frac{D_p v \varrho}{6(1-\varepsilon)\mu} \tag{8.28}$$

In the laminar range, substituting for Re from this equation in eq. (8.24)

$$f_c = \frac{30(1-\varepsilon)\mu}{D_p v \varrho} \tag{8.29}$$

Finally, substituting from this equation for f_c, and from eq. (8.27) for S_p, in eq. (8.22), and simplifying the resulting equation

$$\Delta p = \frac{180\mu h(1-\varepsilon)^2 v}{g_c \varepsilon^3 D_p^2} \tag{8.30}$$

EXAMPLE 8.5

A gas (density $= 0.08$ lb/ft³, viscosity $= 0.04$ lb/(ft)(hr)) is at a steady flow through a bed of spherical particles (voidage fraction $= 0.35$, particle diameter $= 0.2$ in.), at a superficial velocity of 0·1 fps. If the height of the bed is 10 ft, what is the drop in pressure, using:

(a) the Carman–Kozeny equation,
(b) the Chilton–Colburn equation, with unity assigned to the factors F_w and F_p of this equation?

Solution

For spherical particles, from eq. (8.28)

$$Re = \frac{D_p v \varrho}{6(1-\varepsilon)\mu} = \frac{(0.2/12)(0.1)(3600)(0.08)}{6(1-0.35)(0.04)}$$

$$Re = \frac{8}{2.6}$$

This is less than the critical number 4·0, and eq. (8.30) applies.

$$\Delta p = \frac{(180)(0.04)(10)(1-0.35)^2(0.1)(3600)}{(32.2)(3600^2)(0.35)^3(0.2/12)^2}$$

$$\Delta p = 2.20 \text{ psf}$$

Taking $F_w = 1$, $F_p = 1$, in the Chilton–Colburn equation (eq. (8.10))

$$\Delta p = \frac{4f'h\varrho v^2}{2g_c D_p}$$

Using eq. (8.12)

$$N_{Re} = \frac{D_p v \varrho}{\mu} = \frac{(0.2/12)(0.1)(3600)(0.08)}{0.04}$$

$$N_{Re} = 12$$

This is less than the critical value of 40, and eq. (8.13) applies.

$$f' = \frac{850}{N_{Re}} = \frac{850}{12}$$

$$\Delta p = \frac{(4)(850/12)(10)(0.08)(0.1)^2}{(2)(32.2)(0.2/12)}$$

$$\Delta p = 2.11 \text{ psf}$$

EXAMPLE 8.6

A gas (density $= 0.072$ lb/ft^3, viscosity $= 0.05$ lb/(ft)(hr)) flows through a reactor filled with a catalyst to a depth of 4 ft. The catalyst is shaped in the form of cuboids, of sides $0.1 \times 0.1 \times 0.2$ in., and a graduated cylinder of 8 cm diameter filled to a height of 10 cm contains 753·6 gm of these cuboids.

What is the drop in pressure across the bed for a mass velocity of 360 lb/(ft^2)(hr)? The density of the catalyst is 2·5 gm/cm^3.

Solution

$$A_p = 2(0.1)^2 + 4(0.1)(0.2) = 0.1 \text{ in}^2$$

$$V_p = (0.1)^2 (0.2) = 0.002 \text{ in}^3$$

$$\frac{A_p}{V_p} = \frac{0.1}{0.002} = 50 \text{ in}^2/\text{in}^3$$

$$S_p = (50)(12) = 600 \text{ ft}^2/\text{ft}^3$$

Let ε be the voidage fraction of the bed, then from the data

$$(8^2)(\pi/4)(10)(2.5)(1 - \varepsilon) = 753.6$$

from which

$$\varepsilon = 0.4$$

Using eq. (8.21)

$$Re = \frac{G}{S_p(1 - \varepsilon)\mu} = \frac{360}{(600)(1 - 0.4)(0.05)}$$

$$Re = 20$$

For solid particles, and $Re > 4$, eq. (8.25) applies.

$$f_c = \frac{5}{20} + \frac{0.4}{20^{0.1}}$$

$$f_c = 0.55$$

Substituting the data in eq. (8.23)

$$\Delta p = (0.55)\frac{360^2}{(32.2)(3600)^2}\left(\frac{1 - 0.4}{0.4^3}\right)\frac{(4)(600)}{(0.072)}$$

$$\Delta p = 53.4 \text{ psf}$$

EXAMPLE 8.7

A bed of graded sand, of 120 ft² cross-sectional area and 4 ft deep, is used for clarification of water (density = 62·4 lb/ft³, viscosity = 2·4 lb/(ft)(hr)). The bed contains spherical particles of 0·05 in. diameter, and it may be assumed that the effective voidage fraction is 0·2. If the water surface is maintained at a level of 18 ft above the top surface of the sand, what is the rate of flow?

Solution

With reference to the diagram of Fig. 8.7

$$H = \frac{\Delta p}{\gamma} = 18 \text{ ft}$$

where γ is the density of the water, and $\Delta p = p - p_o$.

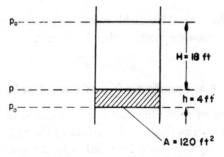

FIG. 8.7. Diagram for Example 8.7.

Since p_o is the atmospheric pressure, then (ignoring the liquid in the pores of the sand) Δp is also the drop in pressure across the sand bed.

Assuming viscous flow through the bed, then taking (from eq. (7.6))

in eq. (8.30)

$$g_c = g\left(\frac{\varrho}{\gamma}\right)$$

and rearranging the terms of the resulting equation

$$\frac{\Delta p}{\gamma} = \frac{180\mu h(1-\varepsilon)^2 v}{g\varepsilon^3 D_p^2 \varrho}$$

Substituting the data in this equation

$$18 = \frac{(180)\,(2{\cdot}4)\,(4)\,(1-0{\cdot}2)^2\,(v)\,(3600)}{(32{\cdot}2)\,(3600^2)\,(0{\cdot}2^3)\,(0{\cdot}05/12)^2\,(62{\cdot}4)}$$

from which

$$v = 0{\cdot}01635 \text{ ft/sec}$$

$$Q = vA = (0{\cdot}01635)\,(120)$$

$$Q = 1{\cdot}962 \text{ ft}^3/\text{sec}$$

Checking on the pattern of flow across the bed

$$S_p = \frac{6}{D_p} = \frac{6}{0{\cdot}05/12} = 1440 \text{ ft}^2/\text{ft}^3$$

$$Re = \frac{v\varrho}{S_p(1-\varepsilon)\mu} = \frac{(0{\cdot}01635)\,(3600)\,(62{\cdot}4)}{(1440)\,(1-0{\cdot}2)\,(2{\cdot}4)}$$

$$Re = 1{\cdot}33$$

This is less than the critical Reynolds number of 4, accepted for the viscous range, and the assumption made earlier is correct.

8.6. Specific Surface of Packed Beds

The surface area of individual particles contained in a unit volume of a packed bed is known as the specific surface of the bed. The usual symbol given to this quantity is a_v, and its relation to the specific surface of the individual particle may be deduced as follows.

Let N_c be the number of individual particles contained in a volume of 1 ft^3 of the packed bed, then by definition

$$a_v = \frac{\text{surface area of all particles}}{1 \text{ ft}^3 \text{ of bed}}$$

$$a_v = \frac{N_c A_p}{N_c V_p(1/1-\varepsilon)}$$

Taking $S_p = (A_p/V_p)$ and simplifying

$$a_v = S_p(1-\varepsilon) \tag{8.31}$$

The Reynolds number (eq. (8.21)) and the Carman–Kozeny equation (eq. (8.23)), when expressed in terms of a_v, assume the forms

$$Re = \frac{G}{a_v \mu} \tag{8.32}$$

$$\Delta p = \left(\frac{f_c}{g_c}\right)\left(\frac{a_v}{\varepsilon^3}\right)\left(\frac{hG^2}{\varrho}\right) \tag{8.33}$$

EXAMPLE 8.8

A gas (density = 0·075 lb/ft³, viscosity = 0·0432 lb/(ft)(hr)) flows through a bed packed with 2 in. nominal size Raschig rings at a mass velocity of 1000 lb/(ft²)(hr). What is the drop in pressure across the bed if its height is 15 ft?

Solution

The 2 in. nominal size Raschig rings have the following characteristics:[27]

Outer diameter	= 2·0 in.
Inner diameter	= 1·5 in.
Height	= 2·0 in.
Voidage fraction	= 0·74
Number per ft³	= 165

Using the above data

$$A_p = \frac{\pi}{4}(2^2 - 1\cdot5^2)(2) + \pi(2 + 1\cdot5)(2)$$

$$A_p = 7\cdot875\pi \text{ in}^2$$

$$a_v = \frac{(165)(7\cdot875)(\pi)}{144}$$

$$a_v = 28\cdot3 \text{ ft}^2/\text{ft}^3$$

Alternatively, since

$$V_p = \frac{\pi}{4}(2^2 - 1\cdot5^2)(2) = 0\cdot875\pi \text{ in}^3$$

$$S_p = \frac{A_p}{V_p} = \frac{7\cdot875\pi}{0\cdot875\pi}(12)$$

$$S_p = 108 \text{ ft}^2/\text{ft}^3$$

Using eq. (8.31)

$$a_v = S_p(1-\varepsilon) = 108\,(1-0\cdot74)$$
$$a_v = 28\cdot1 \text{ ft}^2/\text{ft}^3$$

Taking an average value of $a_v = 28\cdot2$ in eq. (8.32)

$$Re = \frac{G}{a_v\mu} = \frac{1000}{(28\cdot2)\,(0\cdot0432)}$$
$$Re = 821$$

Using eq. (8.26), for hollow particles

$$f_c = \frac{5}{821} + \frac{1}{821^{0\cdot1}} = 0\cdot006 + 0\cdot511$$
$$f_c = 0\cdot517$$

Substituting the data in eq. (8.33)

$$\Delta p = \left(\frac{0\cdot517}{32\cdot2}\right)\left(\frac{28\cdot2}{0\cdot74^3}\right)\frac{(1000/3600)^2\,(15)}{(0\cdot075)}$$
$$\Delta p = 17\cdot24 \text{ psf}$$

8.7. Filtration

Separation of solids from liquids with the aid of a semi-permeable medium, which retains the solids but allows the liquids to pass through, is the usual definition given to filtration. Filtration equipment may be classified in a number of ways, but gravity, pressure and vacuum filters describe adequately, though rather broadly, the type of equipment in its own class.

Some of the terms used in filtration include the filter cake, which is simply a bed of solid particles separated from the feed, called the slurry, and the solid-free filtrate passing through the filter medium. The other terms refer to the cake itself, which is conveniently described as either compressible or incompressible. A cake is said to be incompressible if its voidage fraction, which is here more commonly referred to as porosity, and the cake-forming particles, remain unchanged during filtration. Otherwise the cake is said to be compressible. Although there are no incompressible cakes in practice, most cakes show little compressibility, and may be conveniently treated as incompressible. The latter provides the basis for the theory of filtration, as outlined in the following paragraphs.

8.8. The Theory of Filtration

As the cake grows in thickness during filtration, it offers an increasing resistance to flow. If the resistance of the filter medium is ignored, the differential pressure across the cake may be related to flow by the Carman–Kozeny equation. Let Δp be the required differential pressure to produce a flow at a superficial velocity (v) across a cake of thickness L, then from eq. (8.22)

$$\frac{(\Delta p)g_c}{L} = f_c \left(\frac{1-\varepsilon}{\varepsilon^3} \right) (S_p \varrho v^2) \tag{8.34}$$

Except for a very short period at the beginning of filtration, when the initial layers of solid particles just start forming the cake, the flow is very slow. This gives rise to the generally accepted assertion that viscous conditions invariably exist in filtration. On this basis, eq. (8.24) applies. Substituting in this equation for Re from eq. (8.20)

$$f_c = \frac{5S_p(1-\varepsilon)\mu}{v\varrho} \tag{8.35}$$

Again, substituting for f_c from this equation in eq. (8.34)

$$\frac{(\Delta p)g_c}{L} = \frac{5S_p(1-\varepsilon)\mu}{v\varrho} \left(\frac{1-\varepsilon}{\varepsilon^3} \right) (S_p \varrho v^2)$$

Simplifying

$$\frac{(\Delta p)g_c}{L} = \frac{5S_p^2(1-\varepsilon)^2 v\mu}{\varepsilon^3} \tag{8.36}$$

Let

$$\alpha = \frac{5S_p^2(1-\varepsilon)^2}{\varepsilon^3} \tag{8.37}$$

Since for incompressible cakes, S_p and ε are constants, α is also a constant. This constant may be looked upon as a characteristic property of the cake, and it is sometimes referred to as the permeability of the cake. Introducing this symbol into eq. (8.36)

$$\frac{(\Delta p)g_c}{L} = \alpha\mu v \tag{8.38}$$

Let dV be the differential volume of filtrate collected in a differential time $d\theta$ when the superficial velocity of the filtrate, based on the face area of the cake A, is v, then

$$vA = \frac{dV}{d\theta}$$

or

$$v = \frac{1}{A}\left(\frac{dV}{d\theta}\right) \qquad (8.39)$$

Substituting for v from this equation in eq. (8.38), and rearranging the terms of the resulting equation

$$\frac{1}{A}\left(\frac{dV}{d\theta}\right) = \frac{(\Delta p)g_c}{\alpha\mu L} \qquad (8.40)$$

This equation cannot be integrated unless L is expressed in terms of V. The interrelation is obtained from a material balance as follows.

The volume of cake of a thickness L is LA, and $LA(1-\varepsilon)\varrho_p = $ mass of solids in the cake, where ϱ_p is the density of the cake-forming particles.

Let x be the concentration of solids in the slurry filtered, expressed in pounds per pound of liquid, then the term $x/(1-x)$ represents the solid–liquid mass ratio in the slurry. If V is the volume of filtrate collected when the cake produced has reached its thickness L, and since the term εLA represents the volume of the liquid retained in the pores of the cake, then by a material balance (solids in cake = solids in slurry filtered)

$$LA(1-\varepsilon)\varrho_p = (V+\varepsilon LA)\,(\varrho)\,(x)/(1-x) \qquad (8.41)$$

from which

$$L = \frac{xV\varrho}{A[(1-\varepsilon)\,(1-x)\varrho_p - x\varepsilon\varrho]} \qquad (8.42)$$

where ϱ is the density of the liquid.

Substituting for L from this equation in eq. (8.40)

$$\frac{1}{A}\left(\frac{dV}{d\theta}\right) = \frac{(\Delta p)g_c A[(1-\varepsilon)\,(1-x)\varrho_p - x\varepsilon\varrho]}{\alpha\mu xV\varrho}$$

Rearranging the terms of this equation

$$\frac{dV}{d\theta} = \frac{A^2(\Delta p)\,g_c[(1-\varepsilon)\,(1-x)\varrho_p - x\varepsilon\varrho]}{Vx\alpha\mu\varrho} \qquad (8.43)$$

Let

$$2C_v = \frac{x\alpha\mu\varrho}{[(1-\varepsilon)\,(1-x)\varrho_p - x\varepsilon\varrho]} \qquad (8.44)$$

then

$$\frac{dV}{d\theta} = \frac{A^2(\Delta p)}{2C_v V} \tag{8.45}$$

If μ and ϱ are assumed constants, then for incompressible cakes C_v is also a constant.

Equation (8.45) is the basic rate equation used in filtration problems. Another form may be obtained in terms of cake thickness, as follows.

From eq. (8.41)

$$V = \frac{LA[(1-\varepsilon)(1-x)\varrho_p - x\varepsilon\varrho]}{x\varrho} \tag{8.46}$$

or differentially

$$\frac{dL}{dV} = \frac{x\varrho}{A[(1-\varepsilon)(1-x)\varrho_p - x\varepsilon\varrho)]} \tag{8.47}$$

Also from eq. (8.40)

$$\frac{dV}{d\theta} = \frac{A(\Delta p)g_c}{\alpha\mu L} \tag{8.48}$$

Multiplying each side of this equation by dL/dV

$$\left(\frac{dV}{d\theta}\right)\left(\frac{dL}{dV}\right) = \frac{A(\Delta p)g_c}{\alpha\mu L}\left(\frac{dL}{dV}\right)$$

Cancelling the dV terms on the left-hand side of this equation

$$\frac{dL}{d\theta} = \frac{A(\Delta p)g_c}{\alpha\mu L}\left(\frac{dL}{dV}\right) \tag{8.49}$$

Substituting in this equation for dL/dV from eq. (8.47)

$$\frac{dL}{d\theta} = \frac{A(\Delta p)g_c x\varrho}{\alpha\mu LA[(1-\varepsilon)(1-x)\varrho_p - x\varepsilon\varrho]} \tag{8.50}$$

Letting

$$2C_L = \frac{\alpha\mu[(1-\varepsilon)(1-x)\varrho_p - x\varepsilon\varrho]}{g_c x\varrho} \tag{8.51}$$

$$\frac{dL}{d\theta} = \frac{\Delta p}{2C_L L} \tag{8.52}$$

where C_L is another constant.

The constants C_V and C_L can be evaluated from available data, but it is more usual to determine them experimentally.

The rate equations (8.45) and (8.52) suggest that filtration is carried out at constant pressure throughout. This is approximately correct in some operations, but more commonly the pressure is gradually increased at the beginning of the filtration until it has reached a level suitable for constant-pressure operation. The whole operation may therefore be looked upon as consisting of two steps. It is convenient to assume that during the first step, usually of much shorter duration than the second step, the filtration proceeds at a constant rate. The two steps are shown diagrammatically in Fig. 8.7.

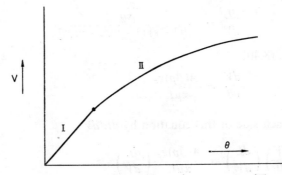

FIG. 8.7a. The constant-rate (I) and constant-pressure (II) steps in a two-stage filtration.

Let subscript o refer to the constant-rate filtration, then we may write

$$\frac{dV}{d\theta} = \frac{V_o}{\theta_o} = \text{constant}$$

where V_o is the volume of filtrate collected in time θ_o, which marks the end of the constant-rate filtration period. At the end of this period, the differential pressure has reached a maximum value (Δp), and from eq. (8.45)

$$\frac{V_o}{\theta_o} = \frac{A^2(\Delta p)}{2C_v V_o}$$

or

$$\theta_o = \frac{2C_v V_o^2}{A^2(\Delta p)} \tag{8.53}$$

If filtration is carried out at a constant pressure from start, eq. (8.45) may be integrated as follows:

$$\int_0^V V(dV) = \frac{A^2(\Delta p)}{2C_v} \int_0^\theta d\theta$$

$$\frac{V^2}{2} = \frac{A^2(\Delta p)\theta}{2C_v}$$

$$\theta = \frac{C_v V^2}{A^2(\Delta p)} \qquad (8.54)$$

where V is the volume of filtrate collected in time θ.

If the constant-pressure filtration is preceded by a constant-rate filtration period, eq. (8.45) may be integrated between the limits V_t, V_o, and θ_t, θ_o, when V_t is the total volume of filtrate collected from the beginning of the filtration in time θ_t. Integrating eq. (8.45) between these limits

$$\int_{V_o}^{V_t} V(dV) = \frac{A^2(\Delta p)}{2C_v} \int_{\theta_o}^{\theta_t} (d\theta)$$

$$\frac{1}{2}(V_t^2 - V_o^2) = \frac{A^2(\Delta p)}{2C_v}(\theta_t - \theta_o)$$

But

$$V_t = V_o + V$$

then

$$V_t^2 - V_o^2 = (V_o + V)^2 - V_o^2 = V^2 + 2V_o V$$

Also taking

$$\theta = \theta_t - \theta_o$$

$$V^2 + 2V_o V = \frac{A^2(\Delta p)\theta}{C_v} \qquad (8.55)$$

Let, in this equation

$$B = \frac{C_v}{A^2(\Delta p)} \qquad (8.56)$$

then

$$\theta = B(V^2 + 2V_o V) \qquad (8.57)$$

Filtration is usually followed by washing. In most types of filtration equipment, the wash liquid follows approximately the path of the filtrate, and washing occurs at the same rate as the rate of filtration at its final stage. This is not the case, however, with a standard filter press, a type of equipment widely used in industry.

Example 8.9

A slurry, containing 0·3 lb of solids (sp.gr. = 3·0) per pound of water, is filtered on a laboratory test filter (see Fig. 8.8) for 4 min at a constant rate. During this period 0·1 ft³ of water is collected per ft² of cake area. The filtration is continued, at the pressure at which the constant-rate filtration was completed, for another 50 min. What is the thickness of the cake produced, if its porosity is 0·5?

Solution

For the constant-rate filtration, $V_o = 0·1$ ft³, $\theta_o = 4$ min, and from eq. (8.53)

$$\frac{A^2(\Delta p)}{C_v} = \frac{2V_o^2}{\theta_o} = \frac{2(0·1)^2}{4}.$$

$$\frac{A^2(\Delta p)}{C_v} = 0·005$$

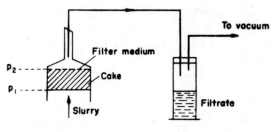

Fig. 8.8. A general arrangement for a filtration test.

or, using eq. (8.56),

$$B = \frac{C_v}{A^2(\Delta p)} = \frac{1}{0·005} = 200$$

where $\Delta p = p_1 - p_2$ (see Fig. 8.8) is the pressure drop across the cake, reached at the end of the constant-rate filtration. This differential pressure is also maintained during the whole constant-pressure filtration period.

For this period, using eq. (8.57)

$$50 = 200[V^2 + (2)(0{\cdot}1)(V)]$$
$$V^2 + 0{\cdot}2V - 0{\cdot}25 = 0$$
$$V = 0{\cdot}41 \text{ ft}^3$$

The total volume of filtrate collected, per square foot of cake area, in 54 min is therefore

$$V_t = V_o + V = 0{\cdot}1 + 0{\cdot}41$$
$$V_t = 0{\cdot}51 \text{ ft}^3$$

The corresponding thickness of the cake produced is obtained from eq. (8.41). Putting in this equation $A = 1$ ft^2, $\varepsilon = 0{\cdot}5$, $x/(1-x) = 0{\cdot}3$, $\varrho_p/\varrho = 3{\cdot}0$, and $V = V_t = 0{\cdot}51$ ft^3

$$L(1)(1-0{\cdot}5)(3{\cdot}0) = [0{\cdot}51 + (0{\cdot}5)(1)(L)](0{\cdot}3)$$

from which

$$L = 0{\cdot}113 \text{ ft}$$

EXAMPLE 8.10

Twenty-four litres of filtrate were collected from a leaf filter in 16 min, under a constant-pressure filtration. What volume of filtrate is produced during the last minute of the filtration?

The cake is to be washed with 3 litres of water at a differential pressure twice as high as that used during the filtration. How long will the washing take?

Solution

Filtration on a leaf filter takes place at a constant pressure throughout, then from eq. (8.54)

$$\frac{A^2(\Delta p)}{C_v} = \frac{V^2}{\theta} = \frac{24^2}{16}$$

$$\frac{A^2(\Delta p)}{C_v} = 36 \frac{(\text{litre})^2}{\text{min}}$$

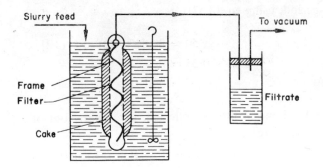

Fɪɢ. 8.9. Filtration on a leaf filter.

From eq. (8.45), the rate of filtration at the end of the filtration period, i.e. during the sixteenth minute of the operation, is

$$\frac{dV}{d\theta} = \frac{A^2(\Delta p)}{C_v}\left(\frac{1}{2V}\right) = \frac{36}{(2)(24)}$$

$$\frac{dV}{d\theta} = 0\cdot75 \text{ litre/min}$$

During the washing, the cake thickness remains unchanged, and the rate of washing is the same as the rate of filtration at its final stage, i.e. 0·75 litre/min, provided that the differential pressure is the same. Doubling this pressure, however, increases the rate of washing to 1·5 litre/min, thus the time to wash with 3 litres of water is 2 min. (It will be noted that the water has been assumed to have the same physical properties as the filtrate.)

EXAMPLE 8.11

A slurry, containing 8 per cent of solids (sp.gr. = 3·0) in water, is filtered on a plate-and-frame press at a constant rate for 6 min. During this period, 0·32 ft³ of filtrate is collected per square foot of cake area. The filtration is then continued at the pressure attained at the end of the constant-rate filtration period. If the cake has a porosity of 0·4, and the frames are 2 in. wide, how long will it take to fill the press?

Solution

For the constant-rate filtration period, from eq. (8.53), taking $A = 1$ ft^2, $V_o = 0.32$ ft^3, and $\theta_o = 6$ min

$$\frac{\Delta p}{C_v} = \frac{2V_o^2}{A^2\theta_o} = \frac{2(0.32)^2}{(1^2)(6)}$$

$$\frac{\Delta p}{C_v} = \frac{(0.32)^2}{3} \text{ ft}^2/\text{min}$$

It will be observed (see Fig. 8.10) that two cakes are produced on each frame, hence a frame 2 in. wide will accommodate two cakes, each of 1 in. thickness. It follows that

$$L = 1/12 \text{ ft}$$

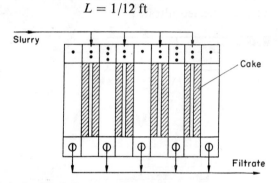

FIG. 8.10. A simplified diagram of flow in a wash-type filter press (. = non-wash plate, .. = frame, ... = wash plate).

and from eq. (8.41), taking $A = 1$ ft^2, $\varrho_p/\varrho = 3.0$, $x/(1-x) = 8/92$

$$\left(\frac{1}{12}\right)(1-0.4)(3.0) = \left[V_t + (0.4)\left(\frac{1}{12}\right)\right]\frac{8}{92}$$

from which, the total volume of the filtrate obtained is

$$V_t = 1.69 \text{ ft}^3$$

The volume of the filtrate collected during the constant-pressure filtration is therefore

$$V = V_t - V_o = 1.69 - 0.32$$
$$V = 1.37 \text{ ft}^3$$

Using eq. (8.55)

$$1\cdot37^2 + 2(0\cdot32)\,(1\cdot37) = \frac{(0\cdot32)^2\theta}{3}$$

from which

$$\theta = 82 \text{ min.}$$

EXAMPLE 8.12

A plate-and-frame press contains twelve frames, each 1 ft×1 ft×1 in. It takes 80 min to fill the press with a given slurry in a constant-pressure filtration. If 0·05 ft³ of cake is produced per 1 ft³ of filtrate, how long will it take to wash the cake with water at the same pressure, using one-tenth of the volume of the collected filtrate?

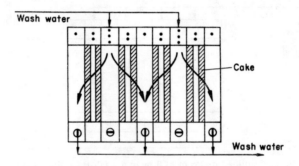

FIG. 8.11. The flow pattern in washing on a wash-type filter press.

Solution

The total volume of the cake produced in the filtration is

$$12\times1\times1\times\frac{1}{12} = 1 \text{ ft}^3$$

The total volume of the filtrate collected is

$$V = \frac{1}{0\cdot05} = 20 \text{ ft}^3$$

The volume of water used in the washing is therefore 2 ft³.

For a constant-pressure filtration, from eq. (8.54)

$$\frac{\Delta p}{C_v} = \frac{V^2}{\theta A^2}$$

Substituting from this equation in eq. (8.45)

$$\frac{dV}{d\theta} = \left(\frac{A^2}{2V}\right)\left(\frac{\Delta p}{C_v}\right) = \left(\frac{A^2}{2V}\right)\left(\frac{V^2}{\theta A^2}\right)$$

$$\frac{dV}{d\theta} = \frac{V}{2\theta} = \frac{20}{(2)(80)} = \frac{1}{8}$$

It follows that at the end of the filtration, the rate of filtration is

$$\frac{dV}{d\theta} = \frac{1}{8} \text{ ft}^3/\text{min}$$

Figure 8.11 shows that the wash liquid follows a different path from that observed in filtration (Fig. 8.10). It flows through two cakes, instead of one, and in addition, the number of feed points in washing is halved when compared with filtration. This pattern of flow demands that the rate of flow of the wash liquid is only one-quarter of the rate of filtration at its final stage, if the differential pressure across the cakes is to be the same. It follows that, in this problem, the washing must proceed at a rate of $\frac{1}{4} \times \frac{1}{8} = \frac{1}{32}$ ft^3/min. With 2 ft^3 of water available, the washing time will therefore be

$$\frac{2 \cdot 0}{1/32} = 64 \text{ min}$$

8.9. Washing in a Centrifuge

Consider, in Fig. 8.12, an instant when a cake formed in a centrifuge is being washed by maintaining the inner surface of the wash liquid at a constant radius r_L.

Let dp be the differential pressure across an elemental thickness of the cake dr at a radius r, when the flow is $dV/d\theta$, then from eq. (8.40)

$$(dp)g_c = -\frac{\alpha\mu(dr)}{A}\left(\frac{dV}{d\theta}\right) \tag{8.58}$$

where the minus sign indicates a decrease in pressure with radius.

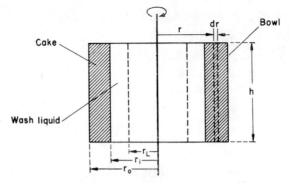

Fig. 8.12. Washing of filter cake in a centrifuge.

The circumferential area of the cake, which is also the flow area for the wash liquid, is related to the radius by

$$A = 2\pi rh \qquad (8.59)$$

where h is the height of the bowl of the centrifuge.

Substituting from this equation for A in eq. (8.58), and rearranging the terms of the resulting equation leads to

$$(dp)g_c = -\frac{\alpha\mu}{2\pi h}\left(\frac{dV}{d\theta}\right)\left(\frac{dr}{r}\right) \qquad (8.60)$$

Integrating

$$g_c \int_{p_o}^{p_i} (dp) = -\frac{\alpha\mu}{2\pi h}\left(\frac{dV}{d\theta}\right)\int_{r_o}^{r_i} \frac{dr}{r}$$

$$(p_i-p_o)g_c = \frac{\alpha\mu}{2\pi h}\left(\frac{dV}{d\theta}\right)\int_{r_i}^{r_o} \frac{dr}{r}$$

Let the pressure drop across the cake be $\Delta p = p_i - p_o$, then

$$(\Delta p)g_c = \frac{\alpha\mu}{2\pi h}\left(\frac{dV}{d\theta}\right)\log_e\frac{r_o}{r_i} \qquad (8.61)$$

In a centrifugal field, the pressure required in filtration is provided by the centrifugal force. Let this force be F, then

$$(dp)g_c = \frac{(dF)g_c}{2\pi rh} \qquad (8.62)$$

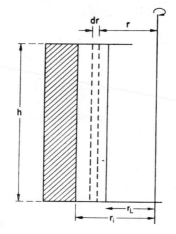

FIG. 8.13. The centrifugal force provided by the wash liquid.

The centrifugal force is the product of the centrifugal acceleration $(\omega^2 r)$, and the mass of the wash liquid contained in the rotating bowl of the centrifuge. Let this mass be m, then with reference to Fig. 8.13

$$dm = 2\pi rh(dr)\varrho \tag{8.63}$$

where ϱ is the density of the wash liquid, and

$$(dF)g_c = 2\pi rh(dr)\,(\omega^2 r)\varrho$$

where ω is the angular velocity of the rotation. Substituting for $(dF)g_c$ from this equation in eq. (8.62), and simplifying

$$(dp)g_c = \varrho\omega^2 r(dr) \tag{8.64}$$

If the mass of the liquid which fills the pores of the cake is ignored, eq. (8.64) can be integrated to give

$$g_c \int_{p_o}^{p_i} (dp) = \varrho\omega^2 \int_{r_L}^{r_i} r\,(dr)$$

$$(\Delta p)g_c = \frac{\varrho\omega^2}{2}\,(r_i^2 - r_L^2) \tag{8.65}$$

From this equation, and eq. (8.61)

$$\frac{\alpha\mu}{2\pi h}\left(\frac{dV}{d\theta}\right)\log_e \frac{r_o}{r_i} = \frac{\varrho\omega^2}{2}\,(r_i^2 - r_L^2)$$

Hence, the rate of washing is given by

$$\frac{dV}{d\theta} = \frac{\pi \varrho \omega^2 h}{\alpha \mu} \left[\frac{r_i^2 - r_L^2}{\log_e (r_o/r_i)} \right] \qquad (8.66)$$

EXAMPLE 8.13

A cake produced in a centrifuge, having a bowl of 18 in. radius and 24 in. high, is to be washed with 5 ft³ of water at 300 rpm, by maintaining its inner surface at a constant radius of 12 in. If the cake is 3 in. thick, how long will the washing take?

The slurry filtered in the centrifuge contained 0·15 lb of solids (sp.gr. = 3·0) per pound of water. In a test on a filter press, of 8 ft² area, the slurry produced 12 ft³ of filtrate in 9 min 10 sec, at a constant pressure of 25 psig.

In each case, the cake may be assumed to have the same porosity of 0·4.

Solution

Filter press. For a constant-pressure filtration, at (25) (144) psfg, from eq. (8.54)

$$C_v = \left(\frac{A}{V} \right)^2 (\Delta p) (\theta) = \left(\frac{8}{12} \right)^2 (25) (144) (550)$$

$$C_v = 0 \cdot 88 \times 10^6 \frac{\text{(lb-f) (sec)}}{\text{ft}^4}$$

From eq. (8.44)

$$\alpha \mu = \frac{2 C_v [(1 - \varepsilon) (1 - x) \varrho_p - x \varepsilon \varrho]}{x \varrho}$$

Taking 1·0 for the specific gravity of the water

$$\alpha \mu = \frac{(2) (0 \cdot 88) (10^6) [(1 - 0 \cdot 4) (1 - 0 \cdot 15) (3) - (0 \cdot 15) (0 \cdot 4) (1 \cdot 0)]}{(0 \cdot 15) (1 \cdot 0)}$$

$$\alpha \mu = 1 \cdot 725 \times 10^7 \frac{\text{(lb-ft) (sec)}}{\text{ft}^4}$$

$$\omega = \frac{2\pi N}{60} = \frac{2\pi (300)}{60}$$

$$\omega = 10\pi \text{ rads/sec}$$

Using eq. (8.66), and taking $h = 2$ ft,

$$r_o = 1 \cdot 5 \text{ ft}, \quad r_i = 1 \cdot 26 \text{ ft}, \quad r_L = 1 \text{ ft},$$

and $\qquad \log_e (r_o/r_i) = 2 \cdot 303 \log (1 \cdot 5/1 \cdot 25) = 0 \cdot 1824$

$$\frac{dV}{d\theta} = \frac{(\pi) (62 \cdot 4) (100\pi^2) (2)}{(1 \cdot 725) (10^7)} \left[\frac{1 \cdot 25^2 - 1 \cdot 0^2}{0 \cdot 1824} \right]$$

$$\frac{dV}{d\theta} = 0 \cdot 0755 \text{ cfs}$$

The time of washing is

$$\frac{5}{0 \cdot 0755} = 66 \text{ sec}$$

8.10. Optimisation in Filtration

A study of the costs involved in the operation of a plant will show that, in most cases, there are certain costs that increase whereas others decrease when related to a common variable. Since the total cost is the sum of all the costs involved, there must be an optimum value of this common variable at which the total cost is a minimum. The process of coordinating the operating variables in order that the total cost be a minimum for a given production, or that a maximum production is attained under specified conditions, is known as optimisation.

Most industrial filtration is carried out intermittently, and filtration itself forms only a part of an operational cycle which usually also encounters a number of unproductive steps. These may include washing, dismantling, cleaning and refitting the filtration plant for the next cycle. The process of coordinating the productive and unproductive steps of a cycle in order that maximum production is attained leads to eq. (8.68), which may be derived as follows.

Consider a two-stage filtration in an intermittently operated plant, such as a filter press, for example. Let V_o and V be the volumes of filtrate collected at constant-rate and constant-pressure filtration periods, respectively, the corresponding times being θ_o and θ. Also let θ_c be the time spent on all the unproductive steps in each filtration cycle then, with reference to Fig. 8.14, the efficiency (η) of the plant is given by

$$\eta = \frac{V_o + V}{\theta_c + \theta_o + \theta} \tag{8.67}$$

The time required for washing, dismantling, cleaning and reassembling the press depends mainly, if not entirely, on the labour force available, and being independent of the filtration itself, may be taken as a constant. The time of the constant-rate filtration is normally fixed arbitrarily by the plant operator and, being also of a short duration, may be taken as a constant as well. It follows that the efficiency of the filtration plant depends solely on the time spent on the constant-pressure filtration. It will be observed (see Fig. 8.14), that the maximum efficiency corresponds to the

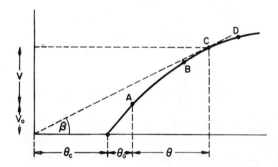

Fig. 8.14. The time–volume relationship in a two-stage filtration.

point C on the constant-pressure curve. The line passing through this point and the origin of the coordinates, i.e. the OC line, is a tangent to this curve, and, the angle β is at its maximum. Any other point on the curve, either below or above the point C, such as the points B or D, when joined to the point O, will produce a line inclined at a smaller angle to the horizontal than the angle β. This angle may then serve as a criterion for the most economical operation of an intermittently operated filtration plant, its relationship to the filtration efficiency being

$$\eta = \tan \beta$$

The maximum efficiency can therefore be determined either graphically[28] from the above relationship, or analytically as follows.

Substituting, in eq. (8.67), for θ, from eq. (8.57)

$$\eta = \frac{V_o + V}{\theta_c + \theta_o + B(V^2 + 2V_oV)}$$

Differentiating this equation with respect to V, and equating to zero for a maximum

$$\frac{d\eta}{dV} = \frac{[\theta_c+\theta_o+B(V^2+2V_oV)]-[(V_o+V)(2BV+2BV_o)]}{[\theta_c+\theta_o+B(V^2+2V_oV)]^2} = 0$$

$$\theta_c+\theta_o+B(V^2+2V_oV) = (V_o+V)(2BV+2BV_o)$$

$$\theta_c+\theta_o = 2BV_oV+BV^2+2BV_o^2$$

Substituting in this equation, from eq. (8.57), for

$$BV^2 = \theta - 2BV_oV$$

and from the equations (8.53) and (8.56) for

$$2BV_o^2 = \theta_o$$

leads to

$$\theta_c = \theta \qquad\qquad (8.68)$$

It follows that, for the most economical operation, the time of constant-pressure filtration is equal to the time during which an intermittently operated plant is out of service. It will be noted, however, that this conclusion has been reached on the assumption that the resistance of the filter medium is negligibly small.

EXAMPLE 8.14

It is intended to carry out a two-stage filtration on a press which takes 20 min to wash, clean and reassemble. The constant-rate filtration is to be completed in 2 min while the pressure builds up to 30 psig. The constant-pressure filtration is to be carried out at this pressure under the most economical conditions. What volume of filtrate will be obtained per cycle under these conditions?

The slurry, when filtered on the same press at a constant pressure of 5 psig, produced 200 litres of filtrate in 4 min 10 sec.

Solution

For the operation at 5 psig, from eq. (8.54), taking $\theta = 25/6$ min

$$\frac{C_v}{A^2} = \frac{\theta(\varDelta p)}{V^2} = \frac{(25/6)(5)}{200^2}$$

$$\frac{C_v}{A^2} = \frac{1}{1920} \frac{(\text{lb-f/in}^2)(\text{min})}{\text{litre}^2}$$

For the constant-rate filtration, substituting the data in eq. (8.53)

$$2 = \frac{2V_o^2}{30} \left(\frac{1}{1920} \right)$$

from which, the volume of the fitrate obtained during this period is

$$V_o = 240 \text{ litres}$$

For the most economical operation, the time of the constant-pressure filtration is, from eq. (8.68),

$$\theta = \theta_c = 20 \text{ min}$$

Substituting in eq. (8.57), from eq. (8.56), for

$$B = \frac{C_v}{A^2(\Delta p)} = \frac{1}{(1920)\,(30)}$$

$$20 = \frac{1}{(1920)\,(30)} \, [V^2 + (2)\,(240)\,(V)]$$

from which

$$V^2 + 480\,V - 1{,}152{,}000 = 0$$

$$V = 758 \text{ litres}$$

The total volume of the filtrate obtained per cycle is then

$$V_t = V_o + V = 240 + 758$$

$$V_t = 998 \text{ litres}$$

8.11. Packed Columns

Vertical shells, filled with a suitable packing material, are widely used in laboratory and industrial practice, mainly in the field of absorption processes. They are popularly called packed columns or towers. The purpose of the filling is to increase the surface of the contact between a gas and liquid, and a number of materials as well as a variety of shapes have been employed or suggested for use in this application. The earliest packings were lump materials such as coke, broken glass or stone, and wooden slats. These are still used occasionally, but lighter manufactured packings, usually in the form of grids, rings or saddles, are the common shapes in use nowadays, the most popular being hollow cylinders marketed under the trade name of Raschig rings.

Pressure losses, accompanying the flow of gases through dry packed columns, may be calculated from the equations derived earlier for beds of solid particles, such as the Carman–Kozeny equation. A straight-line relationship exists in this case between the mass velocity of the gas (G) and the pressure drop (Δp), as shown in Fig. 8.15, which is a logarithmic plot of these variables, with the liquid mass-velocity, as a parameter.

In an absorption column, the free space for the gas flow is markedly reduced by the downflow of the liquid used in the process. This increases

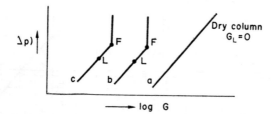

FIG. 8.15. The drop in pressure in dry (a) and wet (b, c) packed columns.

the actual velocity of the gas through the voids of the packing, with the effect that the drop in pressure is considerably higher, for the same mass-velocity of the gas, than it is in a dry column.

For a given liquid flow, there is a gas velocity at which the packed column starts flooding. This condition is described by the flooding points (F) on the diagram, in Fig. 8.15, b, c. The gas velocity at these points is critical, and packed columns are designed to operate below this point, and preferably in the proximity of the so-called loading point, marked on the diagram by a letter L. At these points, the lines on the diagram show a gentle break, although this is usually not distinct enough to be observed. The loading gas velocity corresponds approximately to 60 per cent of the flooding velocity, and this explains why the knowledge of the latter is an important requirement in design practice.

In the absence of experimental data on flooding velocities, use is made of empirical correlations presented by a number of workers. Figure 8.16 shows the most commonly quoted correlation by Lobo et al.[29] Based on an earlier work of Sherwood,[30] the correlation is a logarithmic plot of a dimensionless group of terms, $G_L/G\sqrt{(\varrho/\varrho_L)}$, and a friction factor (f_F) defined by

$$f_F = \frac{G^2 a_v \mu_L^{0.2}}{g_c \varepsilon^3 \varrho \varrho_L} \tag{8.69}$$

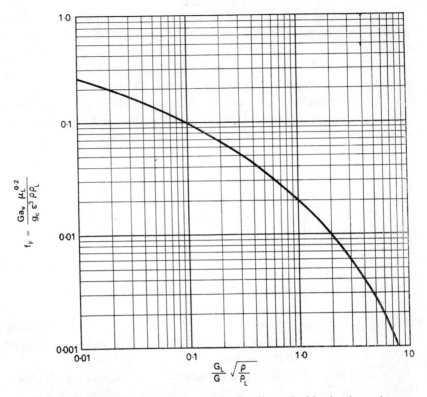

FIG. 8.16. Generalised correlation of the flooding velocities in absorption columns.

where the subscript (L) refers to the liquid. Consistent units are to be used in this equation, except for the liquid viscosity (μ_L) which in this correlation should be expressed in centipoises, when the remaining units are expressed in the fps system.

EXAMPLE 8.15

A column, packed with wooden grids $(a_v = 24 \text{ ft}^2/\text{ft}^3, \ \varepsilon = 0.75)$, is operated at a liquid–gas weight ratio of 2·83. The average densities of the gas and liquid are 0·0382 lb/ft³, and 62·2 lb/ft³, respectively, and the viscosity of the liquid is 0·7 cp. If the gas flows at a mass-velocity of 857 lb/(ft²) (hr), what percentage of the flooding velocity does the flow represent?

Solution

$$\frac{G_L}{G} \sqrt{\frac{\varrho}{\varrho_L}} = 2.83 \sqrt{\frac{0.0382}{62.2}} = 0.07$$

For this value of the dimensionless group, the friction factor (f_F), from Fig. 8.16, is 0·11. Substituting this value, and the data, in eq. (8.69)

$$0.11 = \frac{G^2(24)\,(0.7)^{0.2}}{(32.2)\,(3600)^2\,(0.75^3)\,(0.0382)\,(62.2)}$$

from which the flooding gas velocity is

$$G = 1440 \text{ lb/(ft}^2)\text{ (hr)}$$

The actual mass-velocity of the gas is 857 lb/(ft²) (hr), and this represents

$$\frac{857}{1440} \times 100 = 60 \text{ per cent}$$

of the flooding velocity.

8.12. Pressure Drop in Packed Columns

A number of empirical equations have been suggested for the calculation of the drop in pressure in packed columns. As this drop is some function of both the gas and liquid flow, it is convenient to make use of an equation developed for dry packings, and modify it by introducing a correcting factor to allow for the liquid flow. Figure 8.17 shows how this correcting

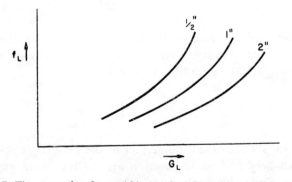

FIG. 8.17. The correcting factor (f_L) as a function of liquid flow (G_L), for three sizes of Raschig rings.

factor (f_L) may be determined from experimentally obtained data,[31] for three sizes of Raschig rings, taken here as an example.

Let $(\Delta p)_w$ be the pressure drop in a packed column, then

$$(\Delta p)_w = f_L(\Delta p) \qquad (8.70)$$

where (Δp) is the drop in pressure, as calculated for dry packing in the same column.

Another approach to the evaluation of the drop in pressure in packed columns is to express this drop in terms of the number of velocity heads

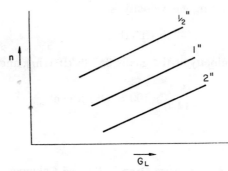

FIG. 8.18. A plot of (n) vs. (G_L), for use with eq. (8.72), for three sizes of Raschig rings.

lost by the gas per foot height of the packed section. Let this number be n, and let the pressure be expressed in engineering units, then

$$\frac{(\Delta p)_w}{h\gamma} = n\left(\frac{v^2}{2g}\right) \qquad (8.71)$$

where h is the height of the packed section, ϱ is the density of the gas, and v is its superficial velocity, based on an empty cross-section. Taking $\gamma = \varrho(g/g_c)$ (eq. (1.6)), eq. (8.71) becomes

$$\frac{(\Delta p)_w}{h\varrho} = n\left(\frac{v^2}{2g_c}\right) \qquad (8.72)$$

Figure 8.18 is a semi-logarithmic plot of the number of velocity heads lost per foot height of packed section against the liquid mass-velocity, for the three sizes of Raschig rings, taken as examples in this diagram. Experimental data for a large number of packings are available.[32]

EXAMPLE 8.16

An absorption column of 6 ft² cross-sectional area is filled with a packing material (S_p = 288 ft²/ft³, ε = 0·4) to a height of 8 ft. When the gas used in the process [density = 0·072 lb/ft³, viscosity = 0·045 lb/(ft) (hr)] flows steadily at a Reynolds number of 100, the observed pressure drop is 20 per cent higher than that calculated for dry packing in this tower, from the Carman–Kozeny equation. If the Carman–Kozeny constant is 0·63, what is the number of velocity heads lost per foot height of the packing? What horsepower is expended in running the fan, with an overall efficiency of 50 per cent?

Solution

From eq. (8.21)

$$G = (Re)\,(S_p)\,(1-\varepsilon)\mu = (100)\,(288)\,(1-0\cdot4)0\cdot045$$
$$G = 777\cdot6\ \text{lb/(ft)}^2\,\text{(hr)}$$
$$v = \frac{G}{3600\varrho} = \frac{777\cdot6}{(3600)\,(0\cdot072)}$$
$$v = 3\cdot0\ \text{fps}$$

Using the Carman–Kozeny equation (eq. (8.22)) for dry packing

$$\frac{\Delta p}{h} = f_c\left(\frac{v^2}{g_c}\right)\left(\frac{1-\varepsilon}{\varepsilon^3}\right)(\varrho S_p)$$
$$= (0\cdot63)\left(\frac{9}{32\cdot2}\right)\left(\frac{1-0\cdot4}{0\cdot4^3}\right)(0\cdot072)\,(288)$$
$$\frac{\Delta p}{h} = 34\cdot22\ \text{psf/ft}$$

Since the observed pressure drop is 20 per cent higher than that calculated from eq. (8.22), then the correcting factor f_L = 1·2, and from eq. (8.70)

$$\frac{(\Delta p)_w}{h} = f_L\frac{(\Delta p)}{h} = (1\cdot2)\,(34\cdot22)$$
$$\frac{(\Delta p)_w}{h} = 41\cdot064\ \text{psf/ft}$$

Substituting this value, and the other data in eq. (8.72)

$$\frac{41\cdot064}{0\cdot072} = n\left(\frac{9}{64\cdot4}\right)$$

from which

$$n = 4080$$

$$(\varDelta p)_w = (41\cdot064)(8) = 328\cdot5 \text{ psf}$$

$$Q = vA = (3)(6) = 18 \text{ ft}^3/\text{sec}$$

$$\text{hp} = \frac{(\varDelta p)_w Q}{(0\cdot5)(550)} \frac{(328\cdot5)(18)}{(0\cdot5)(550)}$$

$$\text{hp} = 21\cdot5$$

8.13. Economic Gas Velocity in Absorption Columns

The economic gas velocity determines the most economical cross-section of an absorption column. The procedure leading to its determination is based on minimising the operating and capital costs, as outlined in the paragraph dealing with optimisation in filtration.

Consider, in Fig. 8.19, a gas and liquid in counterflow, in an absorption column. Let $(\varDelta p)_w$ be the pressure drop across the packed section when the gas flows at a volumetric rate Q, then the theoretical horsepower

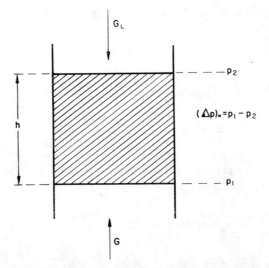

Fig. 8.19. Diagram for the development of eq. (8.82).

expended in driving the gas is given by

$$hp = \frac{Q(\Delta p)_w}{550} \tag{8.73}$$

But $Q = vA$, when A is the cross-sectional area of the column, and v is the superficial velocity of the gas. Expressing the latter in terms of the mass-velocity, on an hourly basis

$$v = \frac{G}{3600\varrho} \tag{8.74}$$

and

$$Q = \frac{GA}{3600\varrho} \tag{8.75}$$

From eq. (8.72)

$$(\Delta p)_w = nh\varrho \frac{v^2}{2g_c}$$

Substituting for v in this equation from eq. (8.74), and putting the numerical value for g_c in the resulting equation

$$(\Delta p)_w = \frac{nhG^2/\varrho}{(64\cdot4)(3600)^2} \tag{8.76}$$

Again, substituting for $(\Delta p)_w$ from this equation, and for Q from eq. (8.75), in eq. (8.73)

$$hp = \frac{nhG^2(GA)}{(64\cdot4)(3600)^2(3600)(550)\varrho^2}$$

Let $V = hA$, be the volume of the packed section of the column, then the last equation may be written

$$hp = 6\cdot05 \times 10^{-16}\frac{nG^3V}{\varrho^2} \tag{8.77}$$

Let θ = operating time, hr/year,
C_e = cost of electricity, pence/kWh,
η = overall efficiency of the fan.

Also, adding 10 per cent to the operational cost for maintenance, and taking $1\,kW = 1\cdot34\,hp$, the annual operating cost is given, in shillings, by

$$C_{op} = \frac{1\cdot1\,hp}{1\cdot34}\left(\frac{\theta}{\eta}\right)\left(\frac{C_e}{12}\right)$$

Substituting in this equation for (hp) from eq. (8.77)

$$C_{op} = \frac{(1 \cdot 1)(6 \cdot 05)(10^{-16})}{(1 \cdot 34)(12)} \left(\frac{\theta}{\eta}\right) \frac{nC_e G^3 V}{\varrho^2}$$

Simplifying

$$C_{op} = 4 \cdot 15 \times 10^{-17} \frac{n\theta C_e G^3 V}{\eta \varrho^2} \tag{8.78}$$

The capital cost must also be related to the packed volume of the column. This volume depends, however, on the so called mass-transfer coefficient, which in turn is a function of G. Let K_g be the overall mass-transfer coefficient, on a gas-phase basis, then for a given absorption process

$$K_g \propto v^a \propto \left(\frac{G}{\varrho}\right)^a$$

where (a) is a constant, characteristic of the process.

Also

$$V \propto \frac{1}{K_g}$$

It follows that

$$V \propto \left(\frac{G}{\varrho}\right)^{-a}$$

or

$$V = k \left(\frac{\varrho}{G}\right)^a \tag{8.79}$$

where k is a proportionality constant.

Let C_i be the investment cost, based on a unit volume of the packed section, and r be the number of years of depreciation of the capital, then the capital cost, expressed in shillings per year, is given by

$$C_{cap} = \frac{VC_i}{r} \tag{8.80}$$

It will be noted that the capital cost includes the interest paid over the r years of depreciation, a period within which the absorption column is supposed to wear out or become totally obsolete.

Let C_T be the total annual cost, in shillings, then

$$C_T = C_{cap} + C_{op}$$

Substituting in this equation for C_{op}, and C_{cap}, from the respective equations (8.78) and (8.80)

$$C_T = V\left[\frac{C_i}{r}+4{\cdot}15\times10^{-17}\frac{n\theta C_e G^3}{\eta\varrho^2}\right] \qquad (8.81)$$

For the gas-film controlling absorption processes, most often met in practice, the characteristic constant (a) of eq. (8.78) is usually taken at a conservative value of 0·8. If we take here $a = 0{\cdot}75$, eq. (8.79) becomes

$$V = k\left(\frac{\varrho}{G}\right)^{0{\cdot}75}$$

and substituting for V from this equation in eq. (8.81)

$$C_T = k\left[\frac{C_i}{r}\left(\frac{\varrho}{G}\right)^{0{\cdot}75}+4{\cdot}15\times10^{-17}\frac{n\theta C_e G^{2{\cdot}25}}{\eta\varrho^{1{\cdot}25}}\right]$$

Differentiating this equation with respect to G, and equating to zero for a minimum cost

$$\frac{1}{k}\left(\frac{dC_T}{dG}\right) = \frac{-0{\cdot}75C_i\varrho^{0{\cdot}75}}{rG^{1{\cdot}75}}+\frac{(2{\cdot}25)\,(4{\cdot}15)\,(10^{-17})n\theta C_e G^{1{\cdot}25}}{\eta\varrho^{1{\cdot}25}} = 0$$

$$G^3 = \frac{(0{\cdot}75)\,(C_i)\,(\eta)\,(\varrho^2)}{(2{\cdot}25)\,(4{\cdot}15)\,(10^{-17})\,(r)\,(C_e)\,(\theta)\,(n)}$$

from which the most economic gas velocity is given by

$$G = 2{\cdot}0\times10^5\left[\left(\frac{C_i}{rC_e}\right)\left(\frac{\eta}{\theta}\right)\left(\frac{\varrho^2}{n}\right)\right]^{1/3} \qquad (8.82)$$

EXAMPLE 8.17

A gas cooling column, packed with wooden grids to a height of 30 ft, is to operate under the conditions listed below. What will be the most economical cross-section of the column, and the horsepower to run the fan under these conditions?

The following data have been extracted from ref. 33:

Gas, flow rate	= 50,000 lb/hr
Gas density	= 0·0382 lb/ft³
Operating time	= 8400 hr/year
Electricity cost	= 1·2 pence/kWh
Capital cost/ft³ packing	= 48 shillings

Depreciation period $= 5$ years
Fan, overall efficiency $= 0.5$
Number of velocity heads
lost per foot height of
the packing $= 9$

Solution

Substituting the data in eq. (8.82), the most economic mass-velocity of the gas is

$$G = 2.0 \times 10^5 \left[\left(\frac{48}{5 \times 1.2} \right) \left(\frac{0.5}{8400} \right) \left(\frac{0.0382^2}{9} \right) \right]^{1/3}$$

$$G = 857 \text{ lb}/(\text{ft}^2)(\text{hr})$$

At the flow of 50,000 lb/hr, the most economic cross-section of the column is therefore

$$A = \frac{50.000}{857} = 58.34 \text{ ft}^2$$

Substituting the data in eq. (8.76)

$$(\Delta p)_w = \frac{(9)(30)(857^2)/0.0382}{(64.4)(3600^2)}$$

$$(\Delta p)_w = 6.22 \text{ psf}$$

The volumetric rate of flow of the gas is

$$Q = \frac{50,000}{(3600)(0.0382)}$$

$$Q = 3636 \text{ ft}^3/\text{sec}$$

At the fan efficiency of 0.5

$$\text{hp} = \frac{(\Delta p)_w Q}{550\eta} = \frac{(6.22)(3636)}{(550)(0.5)}$$

$$\text{hp} = 8.22$$

Problems

8.1. A liquid (density $= 62.0$ lb/ft³, viscosity $= 2.0$ lb/(ft)(hr) flows outside the tubes of a shell-and-tube heat exchanger at a rate of 196,000 lb/hr. The tubes are 16 ft long and of $\frac{3}{4}$ in. outside diameter, and they are arranged in twelve vertical rows on a 1 in.

triangular pitch. The shell has a diameter of $5\frac{1}{4}$ in. and the baffles are spaced at 12 in. intervals. If the drop in pressure across the baffles is 10 per cent of the drop across the tubes, what is the total pressure drop? (*Ans.*: 809 psf)

8.2. Flue gases (mol. wt. = 31) enter a chimney, 100 ft high, at a temperature of 180°C. If the temperature drop is 0·4°C per foot height of the chimney, and the density of the ambient air is 0·0765 lb/ft³, what draught is produced by the chimney?

(*Ans.*: 2·2 psf)

8.3. A chimney is required to produce a draught of 5·6 psf, when flue gases (mol. wt. = 30) enter it at a temperature of 450°F. If the density of the ambient air is 0·076 lb/ft³, and the flue gases are cooled at a rate of 0·5°F per foot height of the chimney, what should be its height? (*Ans.*: 198 ft)

8.4. A catalyst, shaped into cylindrical particles of 0·1 in. diameter and 0·2 in. height, fills a reaction vessel to a height of 5 ft at a voidage fraction of 0·4. What is the specific surface of an individual particle?

If a gas (density = 0·07 lb/ft³), viscosity = 0·042 lb/(ft)(hr) flows through the vessel at a superficial velocity of 0·15 fps, at what Reynolds number does the flow take place, and what is the drop in pressure across the bed at this number?

(*Ans.*: 600 ft²/ft³, 2·5, 2·75 psf)

8.5. A gas (density = 0·0334 lb/ft³, viscosity = 0·0774 lb/(ft)(hr) flows through a bed of solid particles, 4·5 ft deep, at a mass-velocity of 520 lb/(hr) (ft²). The specific area of individual particles is 192 ft²/ft³, and the voidage fraction in the bed is 0·35. If the friction factor is given by eq. (8.26), what is its value, and what is the drop in pressure across the bed? (*Ans.*: 0·763, 195 psf)

8.6. A gas (density = 0·08 lb/ft³, viscosity = 0·05 lb/(ft)(hr)) is at a steady flow through a bed of solid particles ($\varepsilon = 0\cdot4$, $a_v = 800$ ft² per ft³ of the bed) at a mass-velocity of 160 lb/(ft²)(hr). At what Reynolds number (eq. (8.24)) does the flow take place, and what is the drop in pressure across the bed if it is 4·17 ft deep?

(*Ans.*: 4·0, 50 psf)

8.7. A column is filled with dry Raschig rings, of $\frac{1}{4}$ in. nominal size ($\varepsilon = 0\cdot7$, $a_v = 250$ ft²/ft³), to a height of 15 ft. If a gas (density = 0·064 lb/ft³) flows through the column at a superficial velocity of 2 fps, what is the drop in pressure, assuming the Carman–Kozeny friction factor $f_c = 0\cdot6$? If the cross-sectional area of the column is 4 ft², what horsepower is expended in driving the fan at 60 per cent efficiency?

(*Ans.*: 52·14 psf, 1·22 hp)

8.8. A gas (density = 0·07 lb/ft³, viscosity = 0·04 lb/(ft)(hr) flows through a column filled with dry Raschig rings, of a $\frac{1}{2}$ in. nominal size ($\varepsilon = 0\cdot53$, $a_v = 114$ ft²/ft³) to a height of 8 ft. If the gas flows at a mass-velocity of 912 lb/(ft²)(hr), what is the value of the friction factor as calculated from eq. (8.26), and the drop in pressure, using the Carman–Kozeny equation.

Assuming that the Carman–Kozeny equation gives the correct answer, what is the value of the f_p factor in the Chilton–Colburn equation (eq. (8.11)), taking unity for the wall-effect factor? (*Ans.*: 0·615, 107 psf, 0·72)

8.9. A filtration is carried out for 10 min at a constant rate, thereafter it is continued at the pressure attained at the end of the constant-rate filtration. If one-quarter of the total volume of the filtrate is collected during the constant-rate filtration, what is the total filtration time? (*Ans.*: 85 min)

8.10. In a filtration on a laboratory filter, 0·4 ft³ of filtrate were collected in 8 min, at a constant pressure. The cake is to be washed with 0·2 ft³ of water at a pressure half that used in the filtration. Assuming that the water and the filtrate have the same physical properties, how long will the washing take? (*Ans.*: 16 min)

8.11. In a filtration on a standard frame-and-plate press, 24 ft³ of filtrate were obtained in 8 min at a constant pressure of 16 psig. What volume of filtrate was collected during the last minute of the filtration?

The cake produced in the press is to be washed for 2 min with water, at a constant pressure of 4 psig. What volume of water will be required? The water and the filtrate may be assumed to have the same physical properties. (*Ans.*: 1·5 cfm, 0·1875 ft³)

8.12. During filtration on a standard plate-and-frame press, 6 ft³ of filtrate were obtained in 12 min, at a constant pressure of 20 psig. What was the rate of filtration at its final stage?

The cake produced in the press is to be washed with 0·5 ft³ of water at 40 psig. How long will the washing take, assuming the filtrate to have the same physical properties as the wash water? (*Ans.*: 0·25 cfm, 4 min)

8.13. A slurry was filtered on a laboratory filter of 1 ft² area at a constant pressure of 9 psig for 14·4 min. The volume of the filtrate collected during this period was 1·2 ft³.

The same slurry is to be filtered on a plate-and-frame press at a constant pressure of 30 psig for 12 min. What volume of filtrate will be produced per ft² cake area? If the slurry contains 5 weight per cent of solids (sp.gr. = 2·5) in water, and the porosity of the cake is 0·4 in each case, what thickness of cake will be produced in the press? (*Ans.*: 2 ft³, 0·854 in.)

8.14. A slurry, containing 10 weight per cent of solids (sp.gr. = 4·0) in water, was filtered on a plate-and-frame press for 8 min at a constant rate. During this period, 0·2 ft³ of filtrate were obtained, per ft² cake area. The filtration was continued for another 2 hr at the pressure attained at the end of the constant-rate filtration. What was the volume of filtrate obtained per ft² of the cake area during this constant pressure filtration period? If the cake produced had a porosity of 0·6, and the frames were 2 in. wide, what fraction of their volume was filled with the cake at the end of the filtration? (*Ans.*: 0·914 ft³, 0·968)

8.15. A slurry, when filtered on a leaf filter of 1 ft² area at a constant pressure of 8 psig, produced 0·5 ft³ of filtrate in 2½ min.

The same slurry is to be filtered on a plate-and-frame press at a constant pressure of 22·5 psig under the most economical conditions. If dismantling, cleaning and reassembling of the press take 8 min, what will be the volume of the filtrate collected under these conditions, per ft² cake area? If the slurry contains 10 weight per cent of solids (sp.gr. = 2·4) in water, and the cake has a porosity of 0·42 in each case, what will be the thickness of the cake produced? (*Ans.*: 1·5 ft³, 1·5 in. approx.)

8.16. In a test on a plate-and-frame press, a slurry produced 36 ft³ of filtrate in 4 min at a constant pressure of 12 psig.

After cleaning, the press is to be used for filtering the same slurry at a constant pressure of 45 psig, under the most economical conditions. If the cleaning and reassembling of the press take 15 min, what volume of filtrate will be obtained under these conditions? If the slurry contains 10 weight per cent of solids (density = 155 lb/ft³) in water (density = 62·0 lb/ft³), and the cake has a porosity of 0·4 in each case, what will be the total volume of the cake produced, and its weight? (*Ans.*: 135 ft³, 10·3 ft³, 1214 lb)

8.17. In a test on a laboratory filter of 0·8 ft² area, 0·4 ft³ of filtrate were collected in 4 min at a constant pressure of 5 psig.

The test slurry is to be filtered on a plate-and-frame press of 20 ft^2 cake area. It is intended to carry out the filtration at a constant rate for 1 min until 40 psig pressure is reached, and to continue at this pressure under the most economical conditions. If the dismantling, cleaning and reassembling of the press take 24 min, what volume of the filtrate will be obtained in each stage of a filtration cycle? (*Ans.*: 10 ft^3, 60 ft^3)

8.18. A slurry containing 0·2 lb of solids (sp.gr. = 2·5) per pound of water, when filtered on a leaf filter of 2 ft^2 area, at a constant pressure of 10 psig, produced 1 ft^3 of filtrate in 2$\frac{1}{2}$ min.

The same slurry was subsequently filtered in a centrifuge, having a basket of 1·5 ft radius and 2 ft high giving a 6-in.-thick cake. The cake is to be washed with water at 600 rpm, by maintaining its inner surface at a constant radius of 0·5 ft.

If the cake has a porosity of 0·3, what will be the rate of washing under these conditions?
(*Ans.*: 0·247 cfs)

8.19. A gas (density = 0·08 lb/ft^3) flows through an absorption tower (a_v = 24 ft^2/ft^3, ε = 0·6) at a mass-velocity of 720 lb/(hr)(ft^2). If the number of the velocity heads lost per foot height of the wet packing is 480, what is the drop in pressure in a height of 16·1 ft? If the Carman–Kozeny friction factor for the dry packing is f_c = 0·213, what is the value of the f_L factor, as defined by eq. (8.70)? (*Ans.*: 60 psf, 1·407)

8.20. An absorption column, of 5 ft diameter, is packed to a height of 12 ft. If, under the conditions listed below, 500 velocity heads are lost per foot height of the packing, what would be the most economical mass-velocity of the gas used in the absorption process, and the horsepower to run the fan with 50 per cent efficiency, at this velocity?

Capital cost	= 120 sh/ft^3
Gas density	= 0·07 lb/ft^3
Cost of electricity	= 1·2 d/kWh
Operating period	= 8400 hr/year
Depreciation period	= 5 years

(*Ans.*: 453 lb/(ft^2)(hr), 2·76)

CHAPTER 9

PARTICLE DYNAMICS

PARTICLE dynamics is a branch of general mechanics dealing with relative motion between a particle (solid or liquid), and a surrounding fluid (liquid or gas). It is immaterial in this subject whether the particle moves in a stationary fluid or is suspended in a moving fluid, in which case it will appear stationary relative to a solid boundary. It may also be moving countercurrent to or parallel with the fluid, though not necessarily at the same velocity.

The basic theory which follows relates to the motion of a single particle in an infinitely large volume of fluid. Under such an extraneous condition, the particle enjoys complete freedom in its motion, unlike a situation in which a very large number of particles are crowded into a very limited space. If, on the other hand, the concentration of particles is reasonably low, their behaviour may not be very different from that of a single particle. We assume then, that the particles have sufficient freedom in their motion to enable them to behave in nearly the same manner as does the single particle to which the theory applies. We refer to this motion as to *free motion*, generally, or to *free settling*, in particular, when the motion is under the force of gravity. Otherwise the motion is said to be *hindered*. In hindered motion the particles interfere with each other with the effect that their settling rates are reduced considerably.

The theory of settling finds an extensive application in a number of industrially important processes, and the shape of the particle is an important factor in these processes. For the sake of simplicity, however, the particles will be assumed spherical throughout this chapter.

9.1. The Stokes Equation

Consider a solid particle falling from rest in a stationary fluid under the action of gravity. The particle will at first accelerate as it does in a vacuum, but unlike in a vacuum, its acceleration will be retarded due to friction with the surrounding fluid. As the frictional force increases with the velocity, this force will eventually reach a value equal to that of the gravitational force. From this point on, the two forces will be balanced and the particle will continue to fall with constant velocity. Since this velocity is attained at the end of the acceleration period, it is called terminal settling velocity.

In practice, the acceleration period is of a very short duration, often of the order of a small fraction of a second. It is therefore customary to ignore this period in all practical problems concerned with settling processes,

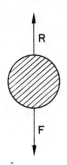

FIG. 9.1. A free falling particle under the action of gravity and resistance forces.

and the terminal settling velocity then becomes the only important factor in this kind of problem. Its magnitude is closely related to the physical properties of the fluid and the particle, and the relationship can be obtained from a balance of the two forces involved.

Consider, in Fig. 9.1, a solid particle of density ϱ_p, falling in a stationary fluid of density ϱ under the action of gravity. The gravitational force (F) acts on the particle even when it is at rest, and remains constant during the whole period of fall. Let Dp be the diameter of the particle, then $(\pi Dp^3/6)$ is its volume, and $(\pi Dp^3\varrho_p/6)$ its mass. From Newton's second law of motion, using the absolute system of units

$$F = \frac{\pi Dp^3\varrho_p}{6}\,(g) - \frac{\pi Dp^3\varrho}{6}\,(g) \qquad (9.1)$$

where g is the gravitational acceleration.

The last term of this equation represents the buoyancy effect. This effect may be ignored if the density of the fluid is negligibly small compared with the density of the solid particle, as is the case when the fluid is a gas. Otherwise eq. (9.1) becomes

$$F = \frac{\pi Dp^3}{6}(\varrho_p - \varrho)g \qquad (9.2)$$

The force resisting motion is generally referred to as the *drag*. Let the drag be designated by the symbol R and let us assume that its magnitude is a function of the diameter of the particle, its velocity u, and of such physical properties of the surrounding fluid as its density and viscosity. On the basis of these assumptions, the following proportionality can be set up

$$R \propto Dp^x u^y \varrho^z \mu^t$$

or

$$R = kDp^x u^y \varrho^z \mu^t \qquad (9.3)$$

where k is a proportionality constant.

Substituting the dimensions for each variable of this equation

$$ML\theta^{-2} = (L)^x (L\theta^{-1})^y (ML^{-3})^z (ML^{-1}\theta^{-1})^t \qquad (9.4)$$

Following Rayleigh's method of dimensional analysis, the dimensions will be separated into the following equations.
Equating the indices for
Mass (M)

$$1 = z + t \qquad \text{(I)}$$

Length (L)

$$1 = x + y - 3z - t \qquad \text{(II)}$$

Time (θ)

$$-2 = -y - t \qquad \text{(III)}$$

Solving the three equations simultaneously in terms of t

$$z = 1 - t \qquad \text{(from I)}$$
$$y = 2 - t \qquad \text{(from III)}$$
$$x = 2 - t \qquad \text{(from II)}$$

Substituting in eq. (9.3) for x, y, and z from the above equations

$$R = k(Dp)^{2-t}(u)^{2-t}(\varrho)^{1-t}(\mu)^t$$

Rearranging the terms

$$R = kDp^2u^2\varrho \left[\frac{\mu}{Dpu\varrho} \right]^t \tag{9.5}$$

This exponential equation does not avail itself for immediate use, except for the condition at which the exponent (t) is clearly determinable. Such a condition exists for laminar flow only. By analogy to flow in pipes, it may be assumed that resistance in laminar motion is inversely proportional to the dimensionless group of terms known as the Reynolds number, defined by

$$Re = \frac{Dpu\varrho}{\mu} \tag{9.6}$$

As in flow through pipes, the Reynolds number serves as a useful criterion for the pattern of motion. For spherical particles, the accepted range for laminar motion is up to $Re = 0.2$, approximately.

Putting $t = 1$, in eq. (9.5), for this range

$$R = kDp^2u^2\varrho \left[\frac{\mu}{Dpu\varrho} \right]$$

Simplifying this equation

$$R = kDpu\mu \tag{9.7}$$

The constant k of this equation has been determined experimentally for a number of shapes. For a spherical particle, it has a value of 3π, so that eq. (9.7) takes the form

$$R = 3\pi Dpu\mu \tag{9.8}$$

This equation was derived in 1851 by Stokes[34] by the application of one of the so-called Navier–Stokes equations to viscous motion in steady flow past a sphere. Equation (9.8) is known as the Stokes equation.

For a particle settling at its terminal velocity the two opposing forces are in a balance, so that

$$F = R$$

Let u_s be the terminal settling velocity, then substituting in the above equation for F and R from the respective equations (9.2) and (9.7)

$$3\pi Dpu_s\mu = \frac{\pi Dp^3}{6} (\varrho_p - \varrho)g$$

from which

$$u_s = \frac{Dp^2(\varrho_p - \varrho)g}{18\mu} \qquad (9.9)$$

Equation (9.9) is also called the Stokes equation.

EXAMPLE 9.1

A particle (sp.gr. = 2·4) is falling in water (sp.gr. = 1, viscosity = 0·01 poise) at a Reynolds number of 0·1. What is the size of the particle and its terminal settling velocity?

Solution

Substituting in eq. (9.6) for $u = u_s$, and taking $Re = 0·1$

$$u_s = \frac{0·1\mu}{Dp\varrho}$$

Again substituting from this equation for u_s in eq. (9.9)

$$\frac{0·1\mu}{Dp\varrho} = \frac{Dp^2(\varrho_p - \varrho)g}{18\mu}$$

from which

$$Dp^3 = \frac{1·8\mu^2}{(\varrho_p - \varrho)g\varrho} = \frac{(1·8)(0·01)^2}{(2·4 - 1·0)(981)(1·0)}$$

$$Dp^3 = 131·1 \times 10^{-9}$$

$Dp = 5·08 \times 10^{-3}$ cm, or 50·8 microns.
    Using eq. (9.9)

$$u_s = \frac{Dp^2(\varrho_p - \varrho)g}{18\mu} = \frac{(5·08 \times 10^{-3})^2(2·4 - 1·0)(981)}{(18)(0·01)}$$

$$u_s = 0·197 \text{ cm/sec}$$

9.2. The Drag Factor

The projected area of a particle is defined as the area of its profile when the particle is in its most stable position. Let this area be A, then for a spherical particle

$$A = Dp^2\pi/4 \tag{9.10}$$

Dividing both sides of eq. (9.5), by this area

$$\frac{R}{A} = \frac{kDp^2}{A} \left[\frac{\mu}{Dpu\varrho} \right]^t u^2\varrho$$

Let

$$C_D = \frac{kDp^2}{A} \left[\frac{\mu}{Dpu\varrho} \right]^t \tag{9.11}$$

then

$$\frac{R}{A} = C_D u^2 \varrho \tag{9.12}$$

Substituting for A in this equation from eq. (9.10)

$$R = Dp^2(\pi/4)C_D u^2 \varrho \tag{9.13}$$

where C_D is called the *drag factor*, or *drag coefficient*.

For a particle settling at its terminal velocity (u_s), $R = F$, and making use of the equation (9.13) and (9.2)

$$Dp^2(\pi/4)C_D u_s^2 \varrho = \frac{\pi Dp^3}{6} (\varrho_p - \varrho)g$$

from which

$$\tfrac{2}{3} Dp(\varrho_p - \varrho)g = C_D u_s^2 \varrho \tag{9.14}$$

or

$$u_s = \sqrt{\left(\frac{2}{3C_D} \right)} \sqrt{\left(\frac{Dp(\varrho_p - \varrho)g}{\varrho} \right)} \tag{9.15}$$

The drag factor is some function of the Reynolds number, as may be expected from eq. (9.11). This number has, therefore, been used in correlating experimental data in the form adopted for the Darcy friction factor. More commonly, however, the correlation involves another factor, of twice the value of C_D. Let this new drag factor be ψ, then

$$\psi = 2C_D \tag{9.16}$$

and eq. (9.15) may be written

$$u_s = \sqrt{\frac{4}{3\psi}} \sqrt{\left(\frac{Dp(\varrho_p - \varrho)g}{\varrho}\right)} \qquad (9.17)$$

The diagram, in Fig. 9.2, is a logarithmic plot of ψ vs. Re, for spherical particles. Three zones can be distinguished on the diagram. The laminar zone, also called the streamline or viscous zone, for Reynolds numbers up

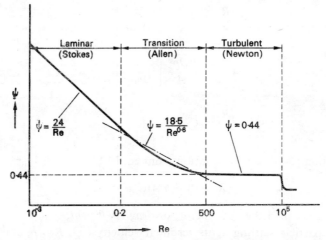

Fig. 9.2. The drag-factor diagram for a spherical particle.

to 0·2, approximately, the transition zone, for Reynolds numbers from 0·2 to 500, approximately, and the turbulent zone, for Reynolds numbers up to about 100,000. The three zones are often referred to as the Stokes, Allen, and Newton zones, respectively.

From the slope of the straight line within the laminar range the following relationship may be obtained:

$$\psi = \frac{24}{Re} \qquad (9.18)$$

Substituting for ψ from this equation, in eq. (9.17), and for Re from eq. (9.6)

$$u_s = \sqrt{\frac{(4)(Dp)(u_s)(\varrho)}{(3)(24)(\mu)}} \sqrt{\frac{Dp(\varrho_s - \varrho)g}{\varrho}}$$

from which

$$u_s = \frac{Dp^2(\varrho_p - \varrho)g}{18\mu}$$

This is the Stokes equation (eg. (9.9)), obtained earlier for the laminar range.

For the turbulent range, $\psi = 0.44$, and is constant between $Re = 500$ and $Re = 100,000$, approximately. Putting this figure for ψ in eq. (9.17)

$$u_s = \sqrt{\left(\frac{4}{(3)(0.44)}\right)} \sqrt{\left(\frac{Dp(\varrho_p - \varrho)g}{\varrho}\right)}$$

$$u_s = 1.73 \sqrt{\left(\frac{Dp(\varrho_p - \varrho)g}{\varrho}\right)} \tag{9.19}$$

which is the equation for the terminal settling velocity of a spherical particle, in the turbulent range.

No regular relationship exists between ψ and Re for the transition range. If the curve, in Fig. 9.2, within this range, can be approximated by a straight line, it may be described by the equation

$$\psi = \frac{18.5}{Re^{0.6}} \tag{9.20}$$

EXAMPLE 9.2

A particle of charcoal (density = 51 lb/ft³) is settling in a gas (density = 0.075 lb/ft³, viscosity = 0.017 cp) at a terminal velocity of 20 ft/sec. What is the diameter of the particle?

Solution

Assuming turbulent range, and substituting the data in eq. (9.19)

$$20 = 1.73 \sqrt{\left(\frac{Dp(51 - 0.075)(32.2)}{0.075}\right)}$$

from which

$$Dp = 0.0061 \text{ ft}$$

Checking on the assumption made

$$Re = \frac{Dpu_s\varrho}{\mu} = \frac{(0\cdot0061)\,(3600)\,(20)\,(0\cdot075)}{(0\cdot017)\,(2\cdot42)}$$

$$Re = 804$$

which is above $Re = 500$, the accepted limiting value for the turbulent range, and correct.

It will be noted that the figure 2·42, which appears in the above equation, is the conversion factor to bring centipoises to the fps system of units, on an hourly basis.

9.3. The ψRe^2 Diagram

The equations used in free settling are often unamenable in solving problems, in which the usual requirement is the terminal settling velocity. This velocity determines the Reynolds number, which in turn defines the range within which one of the three possible settling equations is applicable. The usual procedure is to assume a settling zone first and check the assumption later when calculation has been completed. This is followed by making another assumption if the first one has proved incorrect. To avoid the repetition of the effort, a convenient diagram[44] has been devised. The diagram is a plot of Re vs. ψRe^2 and ψ/Re for the range of Reynolds numbers from 0·1 to 1000, as this range covers the whole transition zone, which is rather difficult to interpret.

The diagram is appended. Its use is restricted to problems with sufficient data to evaluate the ψRe^2 or ψ/Re terms. Their evaluation can be made from the eqns. (9.22) and (9.23), the derivation of which is given below. From eq. (9.17)

$$\psi = \frac{4Dp(\varrho_p - \varrho)g}{3u_s^2\varrho} \tag{9.21}$$

Multiplying ψ by Re^2, and the right-hand side of this equation by its equivalent

$$Re^2 = \left[\frac{Dpu_s\varrho}{\mu}\right]^2$$

$$\psi Re^2 = \frac{4Dp(\varrho_p - \varrho)g}{3u_s^2\varrho}\left[\frac{Dpu_s\varrho}{\mu}\right]^2$$

Simplifying

$$\psi Re^2 = \frac{4Dp^3(\varrho_p - \varrho)g\varrho}{3\mu^2} \qquad (9.22)$$

From this equation, the ψRe^2 term can be evaluated when the terminal settling velocity is an unknown. For the calculated value of the ψRe^2 term, the corresponding value of Re is then read from the diagram. If the Reynolds number falls outside the range of the diagram then either eq. (9.9) (for $Re < 0.1$), or eq. (9.17) (for $Re > 1000$) is used for the calculation of the terminal settling velocity. Otherwise, the terminal settling velocity can be calculated directly from the value of the Reynolds number read from the diagram.

Another useful equation can be obtained for the evaluation of particle diameter when the terminal settling velocity is known.

Dividing the left-hand side of eq. (9.21) by Re and the right-hand side by the group of terms it represents

$$\frac{\psi}{Re} = \frac{4Dp(\varrho_p - \varrho)g}{3u_s^2\varrho} \left[\frac{\mu}{Dpu_s\varrho} \right]$$

Simplifying

$$\frac{\psi}{Re} = \frac{4(\varrho_p - \varrho)g\mu}{3u_s^3\varrho^2} \qquad (9.23)$$

EXAMPLE 9.3

A particle, of 2.7 gm/cm³ density, is settling in water.

(a) If the particle has a diameter of 0.0354 cm, what is its terminal settling velocity?

(b) If the terminal settling velocity of the particle is half that calculated for (a), what is its diameter?

Solution

Substituting the data in eq. (9.22), and taking for water $\varrho = 1$ gm/cm³, $\mu = 0.01$ poise

$$\psi Re^2 = \frac{4(0.0354)^3 (2.7 - 1.0) (981) (1.0)}{3(0.01)^2}$$

$$\psi Re^2 = 987$$

For this value of the ψRe^2 term, $Re = 18$, as read on the diagram, hence

$$\frac{Dpu_s\varrho}{\mu} = 18$$

$$u_s = \frac{(18)(0 \cdot 01)}{(0 \cdot 0354)(1 \cdot 0)}$$

$$u_s = 5 \cdot 086 \text{ cm/sec}$$

For half of this velocity, putting $u_s = 2 \cdot 543$ cm/sec, and the other data in eq. (9.23)

$$\frac{\psi}{Re} = \frac{4(2 \cdot 7 - 1 \cdot 0)(981)(0 \cdot 01)}{(2 \cdot 543)^3 (1 \cdot 0)^2}$$

$$\frac{\psi}{Re} = 1 \cdot 35$$

Again, from the diagram for this number, $Re = 5$, hence

$$\frac{Dpu_s\varrho}{\mu} = 5$$

$$Dp = \frac{5\mu}{u_s\varrho} = \frac{(5)(0 \cdot 01)}{(2 \cdot 543)(1 \cdot 0)}$$

$$Dp = 0 \cdot 0197 \text{ cm}$$

9.4. Hindered Settling

The theory relating to free settling is not directly applicable to suspensions, in which the neighbouring particles interfere with each other with the effect that their settling velocities are considerably reduced. This reduction may be conveniently expressed in terms of the free space available between the particles, as defined by the voidage fraction, and a number of attempts have been made to correlate experimental data on this basis.

Consider uniformly sized particles settling in a tank filled with a liquid of density ϱ and viscosity μ. Let ε be the voidage fraction of the suspension, and u_f be the velocity of the particles relative to a fixed horizontal plane, i.e. their falling velocity, then by analogy to flow through beds of solid particles

$$u_f = \varepsilon u_c \tag{9.24}$$

where u_c is the velocity of the particles relative to the liquid. As this velocity increases with concentration, and the latter has also some effect on the apparent density and viscosity of the suspension, then the equations derived earlier for free settling can be adapted to hindered settling, as settling from concentrated suspensions is called, as follows.

Let ϱ_c be the mean density of the suspension, and μ_c be its apparent viscosity, as defined by

$$\frac{\mu}{\mu_c} = \phi(\varepsilon) \tag{9.25}$$

where $\phi(\varepsilon)$ is some function of ε, then eq. (9.21) may be written in the following modified form:

$$\psi = \frac{4Dp(\varrho_p - \varrho_c)g}{u_c^2 \varrho_c} \tag{9.26}$$

The drag factor (ψ) may be expressed more generally in terms of the Reynolds number (Re) by

$$\psi = \frac{b}{Re^n} \tag{9.27}$$

in which the constants b and n have the following values.

Range	b	n
Laminar (Stokes)	24	1
Intermediate (Allen)	18·5	0·6
Turbulent (Newton)	0·44	0

In hindered settling, the Reynolds number may be presented in the form

$$Re = \frac{Dpu_c \varrho_c}{\mu_c} \tag{9.28}$$

or, substituting in this equation for μ_c from eq. (9.25)

$$Re = \frac{Dpu_c \varrho_c \phi(\varepsilon)}{\mu} \tag{9.29}$$

Now, substituting for Re from this equation in eq. (9.27), and for ψ from the resulting equation, in eq. (9.26)

$$\frac{b\mu^n}{[Dpu_c \varrho_c \phi(\varepsilon)]^n} = \frac{4Dp(\varrho_p - \varrho_c)g}{3u_c^2 \varrho_c}$$

from which

$$u_c^{2-n} = \frac{4Dp^{(1+n)}(\varrho_p - \varrho)\, g[\phi(\varepsilon)]^n}{3b\mu^n \varrho_c^{(1-n)}}$$

or

$$u_c = \left\{ \frac{4Dp^{(1+n)}(\varrho_p - \varrho_c)\, g[\phi(\varepsilon)]^n}{3b\mu^n \varrho_{c}^{1-n}} \right\}^{1/(2-n)} \tag{9.30}$$

Substituting in this equation, for u_c, from eq. (9.24)

$$u_f = \left\{ \frac{4Dp^{(1+n)}(\varrho_p - \varrho_c)\, g\varepsilon^{(2-n)}[\phi(\varepsilon)]^n}{3b\mu^n \varrho_c^{(1-n)}} \right\}^{1/(2-n)} \tag{9.31}$$

No correlations seem to have been presented for the function $\phi(\varepsilon)$, except for the laminar range. For this range, taking $b=24$, and $n=1$, in eq. (9.31)

$$u_f = \frac{4Dp^2(\varrho_p - \varrho_c)\, g\varepsilon[\phi(\varepsilon)]}{(3)(24)(\mu)}$$

Simplifying

$$u_f = \frac{Dp^2(\varrho_p - \varrho_c)g\varepsilon\,[\phi(\varepsilon)]}{18\mu} \tag{9.32}$$

From the experimental data on sedimentation of small uniform particles, Steinour[35] obtained the following correlation for the laminar range only

$$\phi(\varepsilon) = \frac{\varepsilon}{10^{1.82(1-\varepsilon)}} \tag{9.33}$$

For suspensions made up by coarse particles (above 50 microns) the following equation has been suggested for use in hindered settling

$$\frac{u_f}{u_i} = \varepsilon^n \tag{9.34}$$

where u_i is some function of the particle–container diameter ratio. If this ratio is negligibly small, u_i has the same value as the terminal free settling velocity, and eq. (9.34) becomes

$$\frac{u_f}{u_s} = \varepsilon^n \tag{9.35}$$

The relationship has been confirmed experimentally[36] for spherical particles in water. The experimental data were presented in the form of a logarithmic plot of u_f vs. ε, from which the index n was found to have the following values:

$$n = 4 \cdot 6, \quad \text{for} \quad Re < 0 \cdot 2$$
$$n = 2 \cdot 4, \quad \text{for} \quad Re > 500$$

For the intermediate range ($0 \cdot 2 < Re < 500$), this index was found to be some function of Re.

EXAMPLE 9.4

A suspension consists of uniform particles, of 4 gm/cm³ density and 0·015 cm diameter, in a liquid of 1·4 gm/cm³ density and 1·3 cps viscosity. If the volumetric ratio of the particles and the liquid is 1:4, what is their rate of settling?

Solution

One cm³ of the suspension contains 0·2 cm³ of the solid particles and 0·8 cm³ liquid, then the mean density of the suspension is

$$\varrho_c = (0 \cdot 2) \, (4) + (0 \cdot 8) \, (1 \cdot 4)$$
$$\varrho_c = 1 \cdot 92 \text{ gm/cm}^3$$

The voidage fraction of the suspension is

$$\varepsilon = \frac{\text{volume of clear liquid}}{\text{volume of suspension}}$$

$$\varepsilon = \frac{1 \cdot 0 - 0 \cdot 2}{1 \cdot 0} = 0 \cdot 8$$

Assuming laminar range, the term $\varepsilon[\phi(\varepsilon)]$ of eq. (9.32) (using eq. (9.33)), has the value

$$\varepsilon[\phi(\varepsilon)] = \frac{\varepsilon^2}{10^{1 \cdot 82(1 - \varepsilon)}} = \frac{(0 \cdot 8)^2}{10^{1 \cdot 82(1 - 0 \cdot 8)}}$$

$$\varepsilon[\phi(\varepsilon)] = 0 \cdot 285$$

Substituting for this term, and the other data in eq. (9.32)

$$u_f = \frac{(0 \cdot 01)^2 \, (4 \cdot 00 - 1 \cdot 92) \, (981) \, (0 \cdot 285)}{(18) \, (0 \cdot 013)}$$

$$u_f = 0 \cdot 25 \text{ cm/sec}$$

Checking on the Reynolds number, (from eq. (9.24))

$$u_c = \frac{u_f}{\varepsilon} = \frac{0 \cdot 25}{0 \cdot 8}$$

Substituting this value, and taking for $[\phi(\varepsilon)] = 0 \cdot 285/0 \cdot 8$ in eq. (9.29)

$$Re = \frac{(0 \cdot 01)\,(0 \cdot 25)\,(1 \cdot 92)\,(0 \cdot 285)}{(0 \cdot 013)(0 \cdot 8)^2}$$

$$Re = 0 \cdot 16$$

This is less than 0·2, and correct.

Alternatively, assuming that eq. (9.35) applies, and taking the value of the index $n = 4 \cdot 6$, for the laminar range assumed

$$\varepsilon^n = 0 \cdot 8^{4 \cdot 6} = 0 \cdot 36$$

From eq. (9.9)

$$u_s = \frac{(0 \cdot 01)^2\,(4 \cdot 0 - 1 \cdot 4)\,(981)}{(18)\,(0 \cdot 013)}$$

$$u_s = 1 \cdot 09 \ \text{cm/sec}$$

$$u_f = u_s \varepsilon^n = (1 \cdot 09)\,(0 \cdot 36)$$

$$u_f = 0 \cdot 39 \ \text{cm/sec}$$

as compared with 0·25 cm/sec, obtained earlier.

9.5. Classification of Particles

Consider, in Fig. 9.3, a particle in a rising stream of a fluid. Let v be the upward velocity of the fluid, and u_f be the velocity of the particle relative to a fixed horizontal plane, then using the terminal settling velocity (u_s) as a reference, the following limits can be set up:

$v = 0,$ $u_f = u_s$ (particle falling in a stationary fluid at its terminal settling velocity),

$v = u_s,$ $u_f = 0,$ (particle suspended in the fluid),

$v < u_s,$ $u_f > 0,$ (particle settling at a relative velocity $u_f = u_s - v$),

$v > u_s,$ $u_f < 0,$ (particle moving upwards with the fluid, at a relative velocity $u_f = v - u_s$).

The above limits explain the principle involved in the separation of particles of different sizes into fractions of desired characteristics. The process is known as classification.

Particles subjected to classification may vary in density as well as size. The separation of mixtures of this kind is called sorting, as distinct from sizing, which is the name given to classification of particles of the same density.

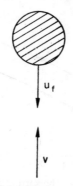

FIG. 9.3. A spherical particle in a rising stream of fluid.

The limits set up for a single particle, or, speaking more generally, for particles at infinite dilution, cannot be applied to suspensions without some refinement. For the sake of simplicity, however, further discussion will refer to very dilute suspensions, when ε is approaching unity, and particles may be assumed to follow the pattern adopted in free settling. On the basis of this simplifying assumption, the process of separating particles into fractions of desired characteristics may be explained with the aid of the so called sorting diagram, as follows.

Consider a very dilute suspension containing a mixture of two different materials, A and B, the material A being the lighter. Let D_{p1} and D_{p2} be the diameters of the smallest and largest particles of the mixture, then plotting terminal settling velocity against particle diameter within the whole range of the sizes available will result in two curves, as shown in Fig. 9.4. Now, assume that classification of the two materials is carried out by a rising stream of a fluid. Two pure fractions can be obtained in this process, by maintaining the upward velocity of the fluid at a predetermined level. Thus, if this velocity is made equal to the settling velocity of the smallest particle $B(v = u_{B1})$ a pure fraction of A will separate from the mixture, as an overhead product. This fraction will contain particles ranging in size between D_{p1} and D_{p3}. Similarly, a pure fraction of B, ranging in size between D_{p4} and D_{p2}, can be obtained as a bottom product (sediment), by maintaining the velocity of the classifying fluid equal to the settling velocity of the largest particle A. The two pure fractions are shown on the diagram (in. Fig. 9.4) by the shaded areas between the curves.

For each fraction, the ratio of the smallest and largest diameters is given by eq. (9.36), which can be obtained as follows.

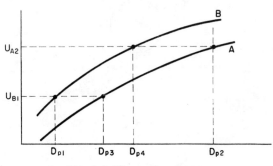

FIG. 9.4. The sorting diagram.

From eq. (9.17), taking $u_A = u_B$, when the subscript A refers to the largest particle A for the pure bottom fraction, or vice versa for the pure overhead fraction

$$u_A^2 = \frac{4(D_{pA})\,(\varrho_{pA} - \varrho)}{3\psi_A \varrho}$$

and

$$u_B^2 = \frac{4(D_{pB})\,(\varrho_{pB} - \varrho)}{3\psi_B \varrho}$$

Equating the above, for the condition $u_A = u_B$, simplifying and rearranging the terms

$$\frac{D_{pA}}{D_{pB}} = \frac{\varrho_{pB} - \varrho}{\varrho_{pA} - \varrho} \times \frac{\psi_A}{\psi_B} \tag{9.36}$$

where the ratio ψ_A/ψ_B is sometimes called the settling coefficient.

In the turbulent range, $\psi_A = \psi_B = 0.44$, eq. (9.36) reduces to

$$\frac{D_{pA}}{D_{pB}} = \frac{\varrho_{pB} - \varrho}{\varrho_{pA} - \varrho} \tag{9.37}$$

In the laminar range

$$\psi = \frac{24}{Re} = \frac{24\mu}{D_p u \varrho}$$

and, since $u_A = u_B$

$$\frac{\psi_A}{\psi_B} = \frac{D_{pB}}{D_{pA}}$$

then, substituting for this ratio from the last equation in eq. (9.36) and rearranging the terms

$$\frac{D_{pA}}{D_{pB}} = \sqrt{\left(\frac{\varrho_{pB} - \varrho}{\varrho_{pA} - \varrho}\right)} \qquad (9.38)$$

Inspection of this equation and of eq. (9.37) shows, at least for the two ranges considered, that—other things being equal—the sorting range can be widened by increasing the density (ϱ) of the classifying fluid. This is occasionally done by adding fine particles, or some reasonably soluble and cheap salts, such as calcium chloride, to the classifying medium.

EXAMPLE 9.5

A mixture of silica (sp.gr. = 2·65) and galena (sp.gr. = 7·5) is to be classified from a very dilute suspension by a rising stream of water. If the largest particles of the mixture are of 0·005 cm in diameter, what velocity of water will give a pure fraction of galena? What will be the smallest diameter of the particles in this fraction?

Solution

With reference to Fig. 9.4a, the rising velocity of water must not fall below the terminal settling velocity (u_{A2}) of the largest particles of silica.
Assuming laminar range

$$v = u_{A2} = \frac{(D_{p2})^2 (\varrho_{pA} - \varrho)g}{18\mu} = \frac{(0·005)^2 (2·65 - 1·0)(981)}{(18)(0·01)}$$

The rising velocity of the water (sp.gr. = 1·0, viscosity = 0·01 poise) is

$$v = 0·225 \text{ cm/sec}$$

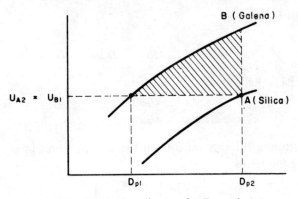

FIG. 9.4a. The sorting diagram for Example 9.5.

Checking on the pattern of flow

$$Re = \frac{(D_{p2})\,(u_{A2})\,(\varrho)}{\mu} = \frac{(0{\cdot}005)\,(0{\cdot}225)\,(1{\cdot}0)}{(0{\cdot}01)}$$

$$Re = 0{\cdot}1125$$

This is less than the critical number of 0·2 for the laminar range, and the assumption made is correct.

The smallest diameter of galena is obtained from the equation

$$u_{B1} = \frac{(D_{p1})^2\,(\varrho_{pB} - \varrho)g}{18\mu}$$

$$D_{p1}^2 = \frac{18 u_{B1}\mu}{(\varrho_{pB} - \varrho)g} = \frac{(18)\,(0{\cdot}225)\,(0{\cdot}01)}{(7{\cdot}5 - 1{\cdot}0)\,(981)}$$

$$D_{p1}^2 = 6{\cdot}35 \times 10^{-6}$$

$$D_{p1} = 0{\cdot}0025 \text{ cm}$$

Alternatively, this diameter may be calculated from eq. (9.38)

$$\frac{0{\cdot}005}{D_{p1}} = \sqrt{\left(\frac{(7{\cdot}5 - 1{\cdot}0)}{(2{\cdot}65 - 1{\cdot}0)}\right)} = 1{\cdot}985$$

$$D_{p1} = \frac{0{\cdot}005}{1{\cdot}985} = 0{\cdot}0025 \text{ cm}$$

9.6. Continuous Sedimentation

Sedimentation may be defined as the process of separation of solid particles from a slurry into a substantially clear liquid and a slurry of a higher concentration of solids, called the sludge. The force causing this separation may be gravitational, centrifugal or of some other form, but the term sedimentation generally implies gravity settling.

If the slurry treated in the sedimentation process contains large amounts of solids, the operation is called *thickening*, otherwise it is known as *clarification*. Alternatively, the operation may be given either name according to whether the thickenend slurry or the overflowing liquid is the desired product.

The mechanism of sedimentation is that of free and hindered settling. It is usually conveniently described in terms of the characteristic zones which appear as the process is in the progress. The work of Coe and Clevenger,[37] published as early as 1916, is still recognised as an important contribution in this field. Their method of correlating the data from small-scale batch settling tests is often utilised in scaling-up operations. Later works[38, 39] may be regarded as useful extensions of this method or merely its refinements. Apart from the references cited, the method will not be described here, and the following discussion will be confined to a brief survey of the settling zones in a continuous thickener and to the basic principles of its design.

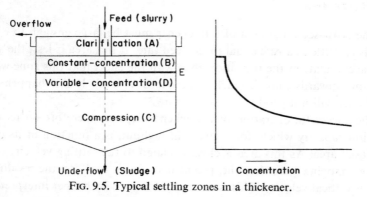

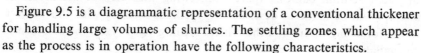

FIG. 9.5. Typical settling zones in a thickener.

Figure 9.5 is a diagrammatic representation of a conventional thickener for handling large volumes of slurries. The settling zones which appear as the process is in operation have the following characteristics.

The clarification zone (A) presents an ideal condition for free settling. Any particle which may have entered this zone with the upward stream of liquid has sufficient time to resettle before the liquid has reached the overflow level.

The constant-concentration zone (B) may also be expected to favour free settling. The concentration of solids in this zone is substantially of the same order as that in the slurry feed.

Between the zone B, and the variable-concentration zone (D), a short transition-zone (E) is distinguished by the lowest settling velocity of the particles. This velocity is called critical, and its determination is essential in design practice. The fact that this critical zone (E) is not to be found at the D–C interface is based on the supposition that the settling velocity of the particles decreases more rapidly than their concentration.

In the compression zone (C), also known as the thickening zone, consolidation of the settled particles takes place. The particles have a limited downward motion, particularly in the upper portion of the zone, as their consolidation is in progress, and this motion produces a slow upward flow of the liquid displaced by the moving particles.

The usual requirement in design practice is the specification of the cross-sectional area and the depth of a thickener. A brief outline of a method leading to this specification is given below.

Thickener Area

The cross-sectional area of a thickener must be large enough to enable solids to settle at a rate equal to the feed rate. If this area is less, the solids will accumulate at the transition zone with the effect that this zone will be moving upwards until it reaches the overflow level. If this happens, the thickener will not produce clear overflow.

The successful operation of a thickener depends therefore on its clarification capacity which, for a given throughput, is a function of its cross-sectional area. As this area is closely related to the settling velocity of the slowest-moving particles, the problem confines itself to the evaluation of this critical velocity. This is normally achieved by proper interpretation of batch settling curves. Apart from citing the relevant literature, no attempt will be made to describe the methods dealing with this subject. In the following discussion the critical settling velocity will be assumed to be available.

Let F = feed rate, lb/hr
  S = sludge, rate of discharge, lb/hr
  O = overflow rate, lb/hr
  X_F = concentration of solids in feed, lb-solids/lb-feed
  X_s = concentration of solids in sludge, lb-solids/lb-sludge

then, on an hourly basis

$$F(1-X_F) = \text{liquid in feed}$$
$$S(1-X_s) = \text{liquid in sludge}$$
$$FX_F \quad = \text{solids in feed}$$
$$SX_s \quad = \text{solids in sludge}$$

Assuming clear overflow, the following liquid balance can be set up

$$O = F(1-X_F)-S(1-X_s) \tag{9.39}$$

Similarly, writing a balance for solids

$$FX_F = SX_s \tag{9.40}$$

Substituting for S from this equation in eq. (9.39), and simplifying

$$O = \frac{F(X_s-X_F)}{X_s} \tag{9.41}$$

Let u = the critical settling velocity
  v = the upward superficial velocity of liquid
  A = the minimum cross-sectional area and
  ϱ = the density of liquid,

then the limiting condition for clear overflow is given by

$$v = u = \frac{O}{A\varrho} \tag{9.42}$$

Substituting in this equation for O from eq. (9.41)

$$A = \frac{F(X_s-X_F)}{(v)(X_s)(\varrho)} \tag{9.43}$$

Thickener Depth

Once the cross-sectional area of the thickener is established, the depth of the compression zone (*C*) can be determined from the retention time of particles in this zone. This time is controlled by the rate of discharge of the sludge which, in turn, depends on the desired concentration in the product. Having fixed this concentration, the volume of the compression zone may be estimated from a consideration of the time each layer of the sediment has been in compression. The estimate is usually made with the aid of test data. The procedure is, however, outside the scope of this text, and—for the sake of simplicity—an average value of this concentration will be assumed to be available in the discussion which follows.

Let X_c be the average concentration, expressed as a mass fraction, in the compression zone, then $\left(\dfrac{1-X_c}{X_c}\right)$ is the average liquid–solid mass ratio in this zone. Now, let θ be the required retention time, in hours, then, since the total volume (V_c) of this zone is the sum of the volumes occupied by the solids and the liquid

$$V_c = \frac{\theta F X_F}{\varrho_p} + \theta F X_F \left(\frac{1-X_c}{X_c}\right)\frac{1}{\varrho}$$

or

$$V_c = \frac{\theta F X_F}{\varrho_p}\left[1+\frac{(1-X_c)}{X_c}\left(\frac{\varrho_p}{\varrho}\right)\right] \tag{9.44}$$

where X_c is some function of θ. Also, let H_c be the required depth of the compression zone, then ignoring the conical section of the thickener

$$H_c = \frac{V_c}{A} \tag{9.45}$$

The total depth of a thickener is estimated approximately by making allowances for the clarification and settling zones, about 2 ft for each, and for the extra storage capacity to cover possible operational irregularities.

EXAMPLE 9.6

A slurry, containing 5 per cent of solids (sp.gr. = 2·6) in water, is to be treated in a thickener at a rate of 16,760 lb/min. The smallest particles to be separated from the slurry have a diameter of 50 microns.

If their critical settling rate is one-quarter of the free settling velocity, and the sludge is to be discharged at a solid concentration of 25 per cent, what should be the minimum cross-sectional area of the thickener?

If the average concentration of the solids in the compression zone is 22 per cent, what will be the depth of this zone at a retention time of 2 hours?

Solution

Assuming viscous conditions in the transition zone of the thickener, the free-settling velocity of the smallest particles is (from eq. (9.9))

$$u_s = \frac{D_p^2(\varrho_p - \varrho)g}{18\mu} = \frac{(0\cdot005)^2\,(2\cdot6-1\cdot0)\,(981)}{(18)\,(0\cdot01)}$$

$$u_s = 0\cdot218 \text{ cm/sec}$$

The critical settling velocity is then

$$u = (0\cdot25)\,(0\cdot218) = 0\cdot0545 \text{ cm/sec}$$

In the fps system of units, taking $v = u$, the upward velocity of the water is

$$v = \frac{0\cdot0545}{30\cdot5} = 0\cdot00179 \text{ fps}$$

From eq. (9.43), the minimum cross-sectional area of the thickener is

$$A = \frac{(16,760)\,(0\cdot25-0\cdot05)}{(60)\,(0\cdot00179)\,(0\cdot25)\,(62\cdot4)}$$

$$A = 1830 \text{ ft}^2$$

Substituting the data in eq. (9.44)

$$V_c = \frac{(2)\,(16760)\,(60)\,(0\cdot05)}{(2\cdot6)\,(62\cdot4)}\left[1+\frac{(1-0\cdot22)}{0\cdot22}\left(\frac{2\cdot6}{1\cdot0}\right)\right]$$

$$V_c = 6333 \text{ ft}^3$$

The depth of the compression zone is then (ignoring the conical section of the thickener)

$$H_c = \frac{V_c}{A} = \frac{6333}{1830}$$

$$H_c = 3\cdot46 \text{ ft}$$

9.7. Centrifugal Clarification

Separation from liquids containing very small amounts of solid material is known as clarification. The very low concentration of particles in clarifiers provides excellent opportunity for free settling, and if, in addition, the particles are very small, laminar motion can be expected even in a centrifugal force field.

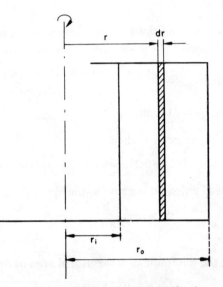

FIG. 9.6. Diagram for centrifugal clarification.

Consider, in Fig. 9.6, a particle suspended in a liquid subjected to a centrifugal force. If the particle is heavier than the liquid, it will travel away from the axis of rotation towards the wall of the rotating bowl which contains the liquid. As centrifugal force varies with the radius, the particle will be at constant acceleration with the effect that it will never reach terminal velocity. In a slow motion, however, we may assume that at any instant the centrifugal force is balanced by the force resisting the motion of the particle. On this basis we may write an equation identical with that developed for the terminal settling velocity in the gravitational field, which is

$$u_s = \frac{D_p^2(\varrho_p - \varrho)g}{18\mu} \tag{9.9}$$

If, in this equation, the gravitational acceleration (g) is replaced by the radial acceleration (a_r), the radial velocity u_r of the particle may be given by

$$u_r = \frac{D_p^2(\varrho_p - \varrho)a_r}{18\mu} \tag{9.46}$$

From the two equations

$$\frac{u_r}{u_s} = \frac{a_r}{g} \tag{9.47}$$

Taking $a_r = \omega^2 r$, where ω is the angular velocity

$$\frac{u_r}{u_s} = \frac{\omega^2 r}{g} \tag{9.48}$$

But the radial particle velocity (u_r) is also given by

$$u_r = \frac{dr}{d\theta}$$

where dr is the elemental radial distance the particle travels in time $d\theta$.

Substituting for u_r from this equation and for u_s from eq. (9.9), in eq. (9.48)

$$\frac{dr}{d\theta} = \left(\frac{\omega^2 r}{g}\right) \frac{D_p^2(\varrho_p - \varrho)g}{18\mu}$$

Rearranging terms

$$\frac{dr}{r} = \frac{D_p^2(\varrho_p - \varrho)\omega^2}{18\mu}(d\theta) \tag{9.49}$$

Let r_o and r_i be the radius of the bowl and that of the inner surface of the liquid, assumed constant, respectively, then integrating eq. (9.49) within these limits

$$\log_e \frac{r_o}{r_i} = \frac{D_p^2(\varrho_p - \varrho)\omega^2\theta}{18\mu} \tag{9.50}$$

This equation is a useful form from which the time (θ) for a particle, to travel the longest possible distance ($r_o - r_i$) can be calculated.

EXAMPLE 9.7

A centrifugal clarifier, with a bowl of radius 60 cm, rotates at a speed of 600 rpm. What time will be required to separate particles of 10 microns in diameter, and of density 2·4 gm/cm³, when the bowl is filled to two-thirds of its volume with the water to be clarified?

Solution

With reference to Fig. 9.6, $r_o = 60$ cm³. The inner radius (r_i) of the liquid surface, when the bowl is in rotation, can be obtained from the equation

$$\tfrac{2}{3} r_o^2 \pi h = \pi (r_o^2 - r_i^2) h$$

from which

$$\frac{r_o}{r_i} = \sqrt{3}$$

At the speed of rotation of $N = 600$ rpm, the angular velocity is

$$\omega = \frac{2\pi N}{60} = 20\pi \text{ radians/sec}$$

Using eq. (9.50)

$$\log_e \sqrt{3} = \frac{(0·001)^2 (2·4 - 1·0) (20\pi)^2 \theta}{(18) (0·01)}$$

from which

$$\theta = 17·9 \text{ sec}$$

Checking on the laminar motion assumed

$$u_r = \frac{D_p^2 (\varrho_p - \varrho) a_r}{18 \mu} = \frac{D_p^2 (\varrho_p - \varrho) \omega^2 r_o}{18 \mu}$$

$$u_r = \frac{(0·001)^2 (2·4 - 1·0) (20\pi)^2 (60)}{(18) (0·01)}$$

$$u_r = 1·845 \text{ cm/sec}$$

$$Re = \frac{D_p u_r \varrho}{\mu} = \frac{(0·001) (1·845) (1)}{0·01}$$

$$Re = 0·1845$$

This is less than 0·2, hence the assumption that the motion is laminar even at the highest value of the radial velocity, when the particle is reaching the wall of the bowl, is correct.

9.8. Separation in Cyclones

The separation of particles from a fluid in cyclones is caused by centrifugal force, as it is in centrifugal clarifiers. This force, however, is not derived from mechanical action but solely from the kinetic energy possessed by the fluid as it enters the cyclone tangentially. This energy makes the particle-laden fluid rotate inside the stationary vertical cylinder, which forms the essential part of a cyclone, and in doing so, it sustains the centrifugal action necessary for the radial motion of the particles. The additional action of the gravitational force provides the downward component of the resultant, and this makes the particles follow a spiral path of increasing radius. If the number of turns, the fluid stream makes before it leaves the cyclone, is sufficient for the particles to reach the wall of the cylinder, they will separate. Otherwise they will be carried away with the stream.

Although the path of particles in cyclones differs from that in centrifugal clarifiers, the principle on which the separation is based is the same for both. Consequently the basic equation developed for centrifugal clarifiers may be adapted to the theory of cyclones.

Consider a particle suspended in a gas stream as it rotates in the cyclone represented diagrammatically in Fig. 9.7. If no slip between the particle

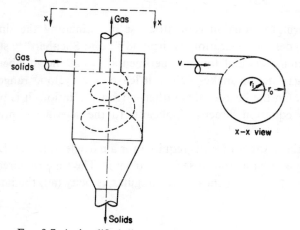

Fig. 9.7. A simplified diagram of a cyclone separator.

and the gas stream may be assumed, then eq. (9.49), developed earlier for the centrifugal clarifier, will apply. Multiplying both sides of this equation by r^2 when r is an instantaneous distance of the particle from the vertical axis of the cyclone, and ignoring the ϱ term as negligibly small for gases, when compared with the ϱ_p term

$$r(dr) = \frac{D_p^2 \varrho_p (\omega r)^2 (d\theta)}{18\mu} \qquad (9.51)$$

Replacing in this equation the ωr term by its equivalent u_T, which is the tangential velocity of the particle

$$r(dr) = \frac{D_p^2 \varrho_p (u_T)^2 (d\theta)}{18\mu} \qquad (9.52)$$

The tangential velocity increases as the particle moves towards the wall. The radial motion, however, is slow and the distance travelled is short, therefore no great error is introduced if this velocity is assumed constant. Equation (9.52) may then be integrated for the extreme case when the particle is supposed to travel across the whole gas stream. The integration gives

$$\frac{r_o^2 - r_i^2}{2} = \frac{D_p^2 u_T^2 \varrho_p \theta}{18\mu} \qquad (9.53)$$

where r_i is the inner radius of the rotating stream, and r_o is the radius of the cylinder.

In this form, the equaion is of little use in estimating the time (θ) required for a desired separation. Stairmand[40] and Shepherd[41] studied the problem extensively and found experimentally that the number of turns which the gas stream makes before it leaves the cyclone range between 0·5 and 3·0. On the basis of this rather narrow limitation, it is possible to develop an equation directly applicable to the separation problems in cyclones.

Let n be the number of turns required for a particle to travel the distance across the gas stream, and so separate from it. The time required to make this number of turns depends on the angular velocity (ω), the interrelation being

$$\theta = \frac{2\pi n}{\omega}$$

This equation may be presented in the form

$$\theta = \frac{2\pi n r_o}{\omega r_o}$$

The denominator of this equation represents the tangential velocity at its highest value, i.e. when the particle has reached the wall. Since this velocity has already been assumed constant, and equal to u_T

$$\theta = \frac{2\pi n r_o}{u_T} \tag{9.54}$$

Substituting for θ from this equation in eq. (9.53)

$$\frac{r_o^2 - r_i^2}{2} = \frac{D_p^2 u_T^2 \varrho_p}{18\mu} \left[\frac{2\pi n r_o}{u_T} \right]$$

Simplifying

$$r_o^2 - r_i^2 = \frac{2\pi D_p^2 u_T r_o n \varrho_p}{9\mu} \tag{9.55}$$

We may further assume that the tangential velocity of the particle has the same value as the average velocity of the gas stream, just before it enters the cyclone. Let this velocity be given by the symbol v, then eq. (9.55) becomes

$$r_o^2 - r_i^2 = \frac{2\pi D_p^2 v r_o n \varrho_p}{9\mu} \tag{9.56}$$

Let the width of the rotating gas stream now be $S = r_o - r_i$, then the left-hand side of eq. (9.56) can be factorised to give

$$r_o^2 - r_i^2 = (r_o - r_i)(r_o + r_i) = S(r_o + r_i)$$

Replacing the radii by the corresponding diameters

$$r_o^2 - r_i^2 = \frac{S(D_o + D_i)}{2}$$

Also, since $D_o + D_i = D_o + (D_o - 2S) = 2(D_o - S)$

$$r_o^2 - r_i^2 = S(D_o - S)$$

Substituting for this in eq. (9.56), and taking $r_o = D_o/2$

$$S(D_o - S) = \frac{\pi D_p^2 v D_o n \varrho_p}{9\mu} \tag{9.57}$$

EXAMPLE 9.8

Particles, of 5 microns diameter and $3 \cdot 0 \ gm/cm^3$ density, are separated in a cyclone from a gas (viscosity $= 0 \cdot 016 \ cp$), when it enters the cylinder at an average velocity of 36 metres per second. What is the probable number of turns made by the gas stream before it leaves the cyclone, if its diameter is $62 \cdot 8 \ cm$, and the width of the gas stream is a quarter of the cyclone's diameter?

Solution

From eq. (9.57)

$$\frac{S(D_o - S)}{D_o} = \frac{\pi D_p^2 v n \varrho_p}{9 \mu}$$

Since $\quad S = \dfrac{D_o}{4}$

$$\frac{S(D_o - S)}{D_o} = \frac{3 D_o}{16}$$

and

$$\frac{3 D_o}{16} = \frac{\pi D_p^2 v n \varrho_p}{9 \mu}$$

Solving this equation for n and substituting the data

$$n = \frac{27 D_o \mu}{16 D_p^2 v \pi \varrho_p} = \frac{(27)\,(62 \cdot 8)\,(0 \cdot 00016)}{(16)\,(0 \cdot 0005)^2\,(3600)\,(3 \cdot 14)\,(3 \cdot 0)}$$

$$n = 2$$

9.9. Fluidisation

If a fluid is passed through a bed of solid particles, the pressure increases with the velocity until a point is reached when the bed starts to expand. At this point the particles lose permanent contact with each other and are free to move throughout the whole bed. The bed then resembles a boiling liquid, and is said to be in a fluidised state.

A typical result for a gas fluidised bed is shown in Fig. 9.8, which is a plot of superficial gas velocity vs. pressure drop, and bed height.

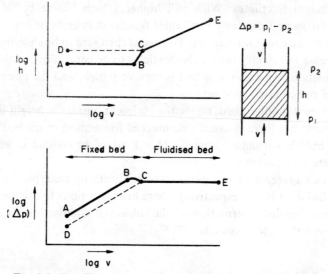

FIG. 9.8. The effect of gas velocity on a bed of solid particles.

The curve (ABC) in Fig. 9.8, is for slowly increasing flow through a bed which has not been fluidised before, the point C being defined as the point of incipient fluidisation. The pressure drop at this point is just sufficient to support the weight of the particles. Let this drop in pressure be Δp, then on the basis of unit cross-section

$$(\Delta p)g_c = h(\varrho_p - \varrho)(1 - \varepsilon)g \tag{9.58}$$

where ε is the porosity of the bed.

At point B the pressure drop is slightly more than enough to support the weight of the particles. The difference between the two pressure drops is due to friction between the individual particles before they can move freely. If the velocity is slowly decreased from point C, the particles are then more loosely packed than they were before fluidisation, and the pressure drop is less for the same gas velocity. Also the bed height h is greater than before.

Below the incipient fluidisation point, the behaviour of the bed is the same for a gas and liquid fluidised bed, but there is a difference in behaviour above that point.

Increasing gas velocity above point C results in the formation of bubbles through the bed which burst on reaching the surface of the bed. This kind of behaviour is called aggregative fluidisation. The bubbles agitate the bed, and its height fluctuates. With still higher velocities, the bubbles grow and appear more frequently until their frontal diameters nearly equal the diameter of the containing vessel. The bed is then said to be slugging. A further increase in velocity causes the particles to be carried out of the vessel. The porosity of the expanding bed is then very high, and the phenomenon is that of pneumatic transport.

For a liquid fluidised bed, above the incipient point, the height increases with velocity but there is usually no marked fluctuation of the bed surface, and no bubble formation occurs. This kind of behaviour is known as particulate fluidisation.

Although aggregative and particulate fluidisations are typical for liquid and gas fluidised beds, respectively, there are exceptions to this generalization. More detailed information on the subject of fluidisation can be obtained from the references cited.[42, 43]

EXAMPLE 9.9

A bed of solid particles (sp.gr. $= 6{\cdot}0$, $\varepsilon = 0{\cdot}4$) is to be fluidised by water. What velocity of flow will produce the state known as incipient fluidisation, if the particles have a diameter of $0{\cdot}001$ ft?

Solution

Using the Carman–Kozeny equation (eq. (8.22)), the drop in pressure across the bed is given by

$$(\Delta p)g_c = f_c v^2 \varrho S_p h \frac{1-\varepsilon}{\varepsilon^3}$$

Substituting for $(\Delta p)g_c$ from this equation, in eq. (9.58), for incipient fluidisation, and simplifying the resulting equation

$$f_c v^2 \varrho S_p = (\varrho_p - \varrho)g\varepsilon^3 \tag{1}$$

Assuming viscous flow, the Carman–Kozeny constant (eq. (8.20) and eq. (8.24)) is given by

$$f_c = \frac{5}{Re} = \frac{5S_p(1-\varepsilon)\mu}{v\varrho}$$

Substituting for f_c from this equation in eq. (1), and simplifying

$$5(1-\varepsilon)S_p^2 v\mu = (\varrho_p - \varrho)g\varepsilon^3$$

from which

$$v = \frac{(\varrho_p - \varrho)g\varepsilon^3}{5(1-\varepsilon)S_p^2\mu} \tag{2}$$

Taking for water $\varrho = 62\cdot4$ lb/ft³, $\mu = 0\cdot000672$ lb/(ft)(sec), and from eq. (8.27), for spherical particles

$$S_p = \frac{6}{D_p} = \frac{6}{0\cdot001} = 6000 \text{ ft}^2/\text{ft}^3$$

$$v = \frac{(6\cdot0 - 1\cdot0)(62\cdot4)(32\cdot2)(0\cdot4)^3}{5(1 - 0\cdot4)(6000)^2(0\cdot000672)}$$

$$v = 0\cdot00886 \text{ fps}$$

Checking on the assumed viscous flow (eq. (8.20))

$$Re = \frac{v\varrho}{S_p(1-\varepsilon)\mu} = \frac{(0\cdot00886)(62\cdot4)}{(6000)(1 - 0\cdot4)(0\cdot000672)}$$

$$Re = 0\cdot23$$

This is less than the critical value of 4, accepted for the viscous range, and proves the validity of the assumption made.

9.10. Accelerated Motion in Free Settling

The period preceding the terminal settling velocity has been ignored so far, on the assumption that it is of a negligibly short duration in the gravitational field. This assumption is quite correct, but it may be of interest to get some idea how short this period is likely to be. Based on earlier works,[44, 45] the following mathematical treatment leads to equations from which this period can be calculated.

Consider, in Fig. 9.9, a particle falling from rest in a fluid, under the force of gravity. As the particle moves, the resistance to its motion increases from zero to a highest value possible, when the particle has reached its terminal settling velocity. Let u be an instantaneous velocity within this range, then its relation to the resistance R is given by eq. (9.13). Replacing in this equation C_D for $\psi/2$

$$R = \frac{\pi D_p^2 \psi u^2 \varrho}{8} \tag{9.59}$$

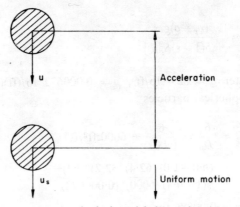

FIG. 9.9. A spherical particle in acceleration.

At any instant, the acceleration is due to the difference between the gravitational force and the force resisting the motion, then from Newton's law, for a particle of mass $\pi D_p^3 \varrho_\varrho / 6$

$$F - R = \frac{\pi D_p^3 \varrho_p}{6} \left(\frac{du}{d\theta} \right) \tag{9.60}$$

Substituting in this equation for F, from eq. (9.2), and for R, from eq. (9.59)

$$\frac{\pi D_p^3 (\varrho_p - \varrho) g}{6} - \frac{\pi D_p^2 \psi u^2 \varrho}{8} = \frac{\pi D_p^3 \varrho_p (du)}{6(d\theta)}$$

from which

$$\frac{du}{d\theta} = \frac{(\varrho_p - \varrho) g}{\varrho_p} - \frac{3u^2 \psi \varrho}{4 D_p \varrho_p} \tag{9.61}$$

From eq. (9.6)

$$u = \frac{\mu(Re)}{D_p \varrho}$$

and

$$du = \frac{\mu}{D_p \varrho} [d(Re)]$$

Substituting for u and du in eq. (9.61)

$$\frac{\mu[d(Re)]}{D_p \varrho (d\theta)} = \frac{(\varrho_p - \varrho) g}{\varrho_p} - \frac{3 \mu^2 (Re)^2 \psi \varrho}{4 D_p^3 \varrho^2 \varrho_p}$$

from which

$$d\theta = \frac{4 D_p^2 \varrho_p \mu [d(Re)]}{4 D_p^3 \varrho (\varrho_p - \varrho) g - 3 Re^2 \psi \mu^2} \tag{9.62}$$

Dividing the numerator and denominator of this equation by $(3\mu^2)$, and simplifying

$$d\theta = \frac{4D_p^2 \varrho_p [d(Re)]/3\mu}{4D_p^3 \varrho(\varrho_p - \varrho)g/3\mu^2 - \psi Re^2} \tag{9.63}$$

Let subscript s refer to the terminal settling velocity, which is a constant for a given situation, then from eq. (9.22)

$$[\psi Re^2]_s = \frac{4D_p^3 \varrho(\varrho_p - \varrho)g}{3\mu^2} \tag{9.64}$$

Substituting from this equation in eq. (9.63)

$$d\theta = \frac{4D_p^2 \varrho_p [d(Re)]/3\mu}{[\psi Re]_s - \psi Re^2} \tag{9.65}$$

$$\theta = \frac{4D_p^2 \varrho_p}{3\mu} \int \frac{d(Re)}{[\psi Re^2]_s - \psi Re^2} \tag{9.66}$$

For the laminar and turbulent range, this equation can be integrated as follows.

(a) Laminar Range

Within this range,

$$\psi = 24/Re$$

then

$$\psi Re^2 = 24 Re$$

and

$$[\psi Re^2]_s = 24(Re)_s$$

Substituting for these terms in eq. (9.66) and integrating between Re, when the particle has reached a velocity u (where $u < u_s$), and $Re = 0$, for the particle at rest

$$\theta = \frac{4D_p^2 \varrho_p}{3\mu} \int_0^{Re} \frac{d(Re)}{[(Re)_s - Re]24}$$

$$\theta = \frac{D_p^2 \varrho_p}{18\mu} \log_e \frac{Re}{(Re)_s - Re} \tag{9.67}$$

Since the Reynolds numbers refer to the same particle and fluid

$$\theta = \frac{D_p^2 \varrho_p}{18\mu} \log_e \frac{u}{u_s - u} \tag{9.68}$$

(b) *Turbulent Range*

Within this range, $\psi_s = \psi = 0.44$, and integrating eq. (9.66) between the limits Re, and $(Re)_o$, which is the Reynolds number when the particle has reached a velocity $u_o < u_s$

$$\theta = \frac{4D_p^2\varrho_p}{(3)\,(0.44)\mu} \int \frac{d(Re)}{(Re)_s^2 - Re^2}$$

$$\theta = \frac{4D_p^2\varrho_p}{(3)\,(0.44)\mu} \left[\frac{1}{2(Re)_s} \right] \log_e \left[\frac{(Re)_s + Re}{(Re)_s - Re} \right]_{(Re)_o}^{Re}$$

Again, as the ratio of the Reynolds numbers equals the ratio of the corresponding velocities, and taking

$$(Re)_s = \frac{D_p u_s \varrho}{\mu}$$

$$\theta = \frac{1.52 D_p^2\varrho_p}{\mu} \left[\frac{\mu}{D_p u_s \varrho} \right] \log_e \left[\frac{u_s + u}{u_s - u} \right]_{u_o}^{u}$$

$$\theta = \frac{1.52 D_p\varrho_p}{u_s\varrho} \log_e \left[\frac{(u_s + u)\,(u_s - u_o)}{(u_s - u)\,(u_s + u_o)} \right] \tag{9.69}$$

EXAMPLE 9.10

A particle, of 0·01 cm diameter and 1·2 gm/cm³ density, falls from rest in water. How long will the particle take to reach 99 per cent of its terminal settling velocity?

Solution

Assuming laminar range, eq. (9.68) may be derived more simply as follows.

Substituting, in eq. (9.60), for F, from eq. (9.2), and for R, from the Stokes resistance equation (eq. (9.8))

$$\frac{\pi D_p^3(\varrho_p - \varrho)g}{6} - 3\pi D_p\mu u = \frac{\pi D_p^3\varrho_p}{6}\left(\frac{du}{d\theta} \right)$$

Multiplying both sides of this equation by $6/\pi D_p'$

$$D_p^2(\varrho_p - \varrho)g - 18\mu u = D_p^2\varrho_p\left(\frac{du}{d\theta} \right)$$

from which

$$d\theta = \frac{D_p^2 \varrho_p (du)}{D_p^2 (\varrho_p - \varrho) g - 18 \mu u}$$

This equation may be presented in the form

$$d\theta = \frac{D_p^2 \varrho_p (du)}{\left[\dfrac{D_p^2 (\varrho_p - \varrho) g}{18 \mu} - u \right] (18 \mu)}$$

Introducing the terminal settling velocity (u_s) from eq. (9.9)

$$d\theta = \frac{D_p^2 \varrho_p (du)}{18 \mu (u_s - u)}$$

Integrating this equation between the limits $u = u$, and $u = 0$, when the particle is at rest

$$\theta = \frac{D_p^2 \varrho_p}{18 \mu} \log_e \left[\frac{u}{u_s - u} \right] \qquad (9.68)$$

For $\qquad u = 0.99 u_s$

$$\log_e \left[\frac{u}{u_s - u} \right] = \log_e \left[\frac{0.99 u_s}{u_s - 0.99 u_s} \right] = \log_e 99$$

$$\log_e \left[\frac{u}{u_s - u} \right] = 4.60$$

Using eq. (9.68)

$$\theta = \frac{(0.01)^2 (1.2) (4.60)}{(18) (0.01)}$$

$$\theta = 0.00307 \text{ sec}$$

Checking on the pattern of motion

$$u_s = \frac{D_p^2 (\varrho_p - \varrho) g}{18 \mu}$$

$$Re = \frac{D_p u_s \varrho}{\mu} = \frac{D_p^3 (\varrho_p - \varrho) \varrho g}{18 \mu^2}$$

$$Re = \frac{(0.01)^3 (1.2 - 1.0) (1.0) (981)}{(18) (0.01)^2}$$

$$Re = 0.109$$

This is less than 0.2, the accepted limit for the laminar motion assumed.

Problems

(In the following problems, the density and viscosity of water should be taken as 1 gm/cm³, and 0·01 poise, respectively.)

9.1. A steel ball (sp.gr. = 7·9) is settling in an oil, of specific gravity 0·95, at a terminal settling velocity of 6 cm/sec. If the ball has a diameter of 0·5 cm, what is the viscosity of the oil? What is the Reynolds number corresponding to this velocity?

(*Ans.*: 15·78 poise, 0·18)

9.2. A particle of 0·01 cm diameter, and 2·4 gm/cm³ density, is settling in an oil of 0·9 gm/cm³ density at a Reynolds number of 0·1. What is the viscosity of the oil? What is the terminal settling velocity at this Reynolds number? (*Ans.*: 0·027 poise, 0·30 cm/sec)

9.3. A particle, of specific gravity 1·2, is settling in water at a Reynolds number of 500. What is the diameter of the particle? (*Ans.*: 0·349 cm)

9.4. A uniform suspension is prepared by 1200 gm of solid particles stirred into enough water to make 1200 cm³ of the suspension. If the particles are of 0·01 cm diameter, and of specific gravity 2·4, what is the mean density of the suspension, and its voidage fraction? What is their rate of falling, assuming that eq. (9.33) applies? What is the corresponding Reynolds number? (*Ans.*: 19/12 gm/cm³, 7/12, 0·0264 cm/sec, 0·0077)

9.5. A suspension is made up of silica particles (diameter = 0·005 cm, density = 2·7 gm/cm³) in water. If the particle–water mass ratio in the suspension is 1 : 2, what is its mean density, and the voidage fraction? Assuming that eq. (9.33) applies, what is the falling velocity of the particles? (*Ans.*: 1·27 gm/cm³, 0·843, 0·072 cm/sec)

9.6. A mixture of silica (sp.gr. = 2·6) and galena (sp.gr. = 7·5) is to be classified from a very dilute suspension with benzene (sp.gr. = 0·85, viscosity = 0·65 cp). If the smallest particles of the mixture are of 0·0005 cm in diameter, what upstream velocity of benzene will produce a maximum amount of pure silica? What will be the largest particle size of the pure silica fraction? Stokes' law may be assumed to apply.

(*Ans.*: 0·014 cm/sec, 0·00096 cm)

9.7. A continuous thickener, of 228 ft² cross-sectional area, treats a slurry at a rate of 124·8 lb/min. The slurry contains 2 per cent of solids (sp.gr. = 2·8, the smallest diameter = 0·001) in water, and the sludge is discharged at a concentration of 22 per cent. If the critical velocity, based on the smallest particles, is given by

$$u = ku_s$$

where u_s is the terminal free settling velocity, what is the value of the constant k? Stokes' law may be assumed to apply. (*Ans.*: 0·413)

9.8. Particles of 5 microns in diameter, and of 2·6 gm/cm³ density are to be separated from water in a centrifugal clarifier. If the separation is to be completed in 3 min when the bowl of the clarifier is one-quarter full, what speed of rotation will be required? If the bowl has a diameter of 50 cm, what is the radial velocity of the particles as they are approaching the wall, and what is the corresponding Reynolds number? Stokes' law may be assumed to apply. (*Ans.*: 181 rpm, 0·02 cm/sec, 0·01)

9.9. A particle, of 0·1 cm diameter and of 1·1 gm/cm³ density, is falling from rest in an oil of 0·9 gm/cm³ density, and 0·3 poise viscosity. Assuming that Stokes' law applies, how long will the particle take to reach 99 per cent of its terminal settling velocity? What is the Reynolds number corresponding to this velocity? (*Ans.*: 0·0093 sec, 0·109)

THE ΨRe^2 DIAGRAM

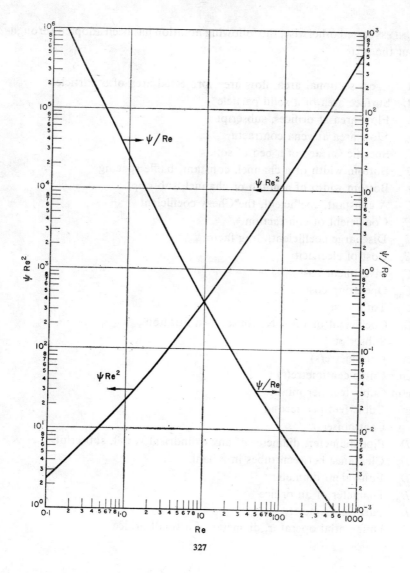

NOTATION

U<small>NLESS</small> stated otherwise, the following notation has been adopted throughout the text:

A Cross-sectional area, flow area, projected area of a particle
A_p Surface area of a solid particle
a Flow area at orifices, subscript
a_c Flow area at vena contracta
a_v Specific surface of a bed of solids
B Bottom width of a channel, constant, baffle spacing
b Bottom width of a notch or channel, subscript
C A constant, coefficient, the Chèzy coefficient
C_c Coefficient of contraction
C_D Discharge coefficient, drag factor
C_e Cost of electricity
C_i Investment cost
C_{op} Operating cost
C_T Total cost
C_v Coefficient of velocity, constant in fitration
c Subscript
cm Centimetre(s)
cm³ Cubic centimetre(s)
cfm Cubic feet per minute
cfs Cubic feet per second
cp   Centipoise(s)
D Pipe diameter, diameter of any cylindrical vessel, subscript
D_c Clearance between tubes in a bank
D_e Equivalent diameter
D_o Diameter of an orifice
D_p Particle diameter
d Differential operator, diameter of a small orifice

E Energy, bulk modulus of elasticity

e Base of natural logarithm

e Absolute roughness of pipes, manometric efficiency, subscript

F Force

f Darcy friction factor, subscript

f' Moody friction factor ($f' = 4f$)

f_c Carman friction factor

f_F Friction factor for packed columns

ft   Foot (unit of length), feet

fps   Feet per second

G Mass-velocity ($G = v\varrho$)

g Gravitational acceleration, subscript

g_c Gravitational constant (conversion factor for force)

gm   Gram(s)

H Fluid head, pressure head

H_f Total loss of energy (head)

H_m Manometric head

H_p Rise in pressure head through an impeller

H_s Specific head in open channels

H_v Velocity head

h Height, fluid head

h_c Loss in sudden contraction

h_e Loss in sudden expansion

h_f Loss due to friction in pipes

hp   Horsepower

hr   Hour(s)

i Slope of channel bed, subscript

in.   Inch(es)

j Subscript

K A constant

k Ratio of specific heats, proportionality constant

K_g Overall mass-transfer coefficient

kg   Kilogram(s)

L Length, dimension of length

l Linear dimension

lb   Pound(s)

lb-f   Pound force (1 lb-f = 32·2 poundals)

M Dimension of mass, Mach number

M_w Molecular weight

m Metre(s)

m Mass, mean hydraulic depth in open channels

min Minute(s)

N Number of revolutions per minute, a number

N_{Re} Reynolds number for Chilton–Colburn equation

n Exponent, a number of

O Rate of overflow

o Subscript

P Power, wetted perimeter

P_t A pitch (transverse distance between centres of adjacent tubes)

p Pressure, subscript

psf Pound-force per square foot

psi Pound-force per square inch

Δp Differential pressure, pressure drop

Q Volumetric rate of flow

R Universal gas constant, resistance force

R' Resistance force per unit surface area

Re Reynolds number

R_H Hydraulic radius

r Radius, radial distance, years of depreciation, subscript

rpm Revolutions per minute

S Surface area, width of a rotating fluid stream

S_p Specific surface of a solid particle

s Subscript

sec Second(s)

T Absolute temperature, dimension of temperature

t Temperature, subscript

u Velocity of a solid body or of a specific part of the body, relative to fluid

u_r Radial velocity

u_s Terminal settling velocity

u_T Tangential velocity

V Volume, specific volume

V_m Volume of one gram-molecule, or pound-molecule of a gas at STP

V_p Volume of a solid particle

v Average fluid velocity (superficial)

$\bar{v}$ Average fluid velocity through voids of a solid bed

v_c velocity at vena contracta

$v_{\max}$ Maximum velocity

v_s Sonic velocity

v_x Point velocity at a radial distance (y)

W Work, weight

w Subscript

x A fraction of unity

X Mass-ratio

y Radial distance

Z Vertical distance above reference line or plane

Greek letters

α (alpha)	A factor, correcting factor for kinetic energy, a filtration constant, angle
β (beta)	Orifice-pipe diameter ratio, angle
γ (gamma)	Weight density
Δ (delta)	The difference in value
ε (epsilon)	Voidage fraction, porosity
η (eta)	Efficiency
Θ (theta)	Time, dimension of time, angle
μ (mu)	Viscosity (dynamic)
ν (nu)	Kinematic viscosity
π (pi)	Circumference–diameter ratio of circles, 3·14
ϱ (rho)	Mass density of a fluid
ϱ_h	Density of hot chimney gas
ϱ_L	Density of a liquid
τ (tau)	Shear stress
ϕ (phi)	A function of, angle
ψ (psi)	Drag factor for solid particles ($= 2C_D$)
ω (omega)	Angular velocity

REFERENCES

1. ROUSE, H. and INCE, S. (1957) *History of Hydraulics*, Dover Publications Inc., New York.
2. SCHLICHTING, H. (1955) *Boundary Layer Theory*, Pergamon Press.
3. TOWNSEND, A. A. (1956) *The Structure of Turbulent Shear Flow*, Cambridge University Press.
4. FOCKEN, C. M. and DINGLE, H. (1933) *Dimensional Methods and their Applications*, Edward Arnold & Co.
5. LINFORD, A. (1961) *Flow Measurement and Meters*, E. & F. N. Spon Ltd., London.
6. COLEMAN M. C. (1956) *Variable area flowmeters*, Trans. Inst. Chem. Eng. **34** (4).
7. NIKURADSE, J. (1953) *V.D.I. Forschungshefte*, 361.
8. MOODY, L. F. (1944) *Trans. Asme* **66**, 671–84.
9. ROUSE, H. (1950) *Engineering Hydraulics*, Wiley, New York.
10. CHOW, V. T. (1959) *Open-Channel Hydraulics*, McGraw-Hill.
11. SHAPIRO, A. H. (1953) *The Dynamics and Thermodynamics of Compressible Fluid Flow*, The Ronald Press Company, New York.
12. CAMBEL, A. B. and JENNINGS, B. H. (1958) *Gas Dynamics*, McGraw-Hill.
13. GEYER, E. W. (1939) *The Engineer*, **168**, 4358, 60–63.
14. GASTERSTADT, VOGT and WHITE (1948) *Ind. Eng. Chem.* **40**, 425.
15. MICHELL, S. J. (1957) Design of pneumatic conveyors, *Brit. Chem. Eng.* **2**, 356.
16. MICHELL, S. J. (1965) Pneumatic conveying, *Loughborough University Chem. Eng. Journal*, **1**, 27.
17. PERRY, J. H. (1950) *Chemical Engineer's Handbook*, 3rd ed., McGraw-Hill.
18. KERN, D. Q. (1950) *Process Heat Transfer*, McGraw-Hill.
19. MICHELL, S. J. (1957) Natural and balanced draft systems, *British Chemical Engineering*, **2** (11), 590.
20. DARCY, H. P. G. (1856) *Les Fontaines publiques de la ville de Dijou*, Paris.
21. COULSON, J. M. and RICHARDSON, J. P. (1955) *Chemical Engineering*, Pergamon Press.
22. SHERWOOD, T. K. and PIGFORD, R. L. (1952) *Absorption and Extraction*, McGraw-Hill.
23. FURNAS, C. C. (1929) *U.S. Bur. Mines Bull.* 307.
24. WHITE, A. M. (1942) *Trans. Amer. Inst. Chem. Eng.* **31**, 390.
25. KOZENY, J. (1927) Uber kapillare Leitung der Wasser Boden, *Ber. Wien Akad.* **136 A**, 271.
26. CARMAN, P. C. (1937) Fluid flow through granular beds, *Trans. Inst. Chem. Eng.* **15**, 150.
27. NORMAN, W. S. (1961) *Absorption, Distillation and Cooling Towers*, Longmans, London.
28. MICHELL, S. J. (1968) Optimisation in filtration, *Technika—Nauka* (30), London.

29. LOBO, W. E., FRIEND, L., HASHMALL, F. and ZENZ, F. (1945) *Trans. Amer. Inst. Chem. Eng.* **41**, 692.
30. SHERWOOD, T. K., SHIPLEY, G. H. and HOLLOWAY, F. A. L. (1938) *Ind. Eng. Chem.* **30**, 765.
31. TILLSON, P. (1939) M.Sc. Thesis, M.I.T., U.S.A.
32. MORRIS, G. A. and JACKSON, J. (1953) *Absorption Towers*, Butterworths Scientific Publications, London.
33. MICHELL, S. J. (1958) Designing a gas cooling tower, *Brit. Chem. Eng.* **3** (7), 372.
34. STOKES, G. G. (1851) *Trans. Cambridge Phil. Soc.* **9**, 51.
35. STEINOUR, R. H. (1944) *Ind. Eng. Chem.* **36**, 618, 840, 901.
36. RICHARDSON, J. F. and ZAKI, W. N. (1954) *Trans. Inst. Chem. Eng.* **32**, 35.
37. COE, H. S. and CLEVENGER, G. H. (1916) *Trans. Amer. Inst. Min. Eng.* **55**, 356.
38. KYNCH, G. J. (1952) *Trans. Faraday Soc.* **48**, 166.
39. ROBINS, W. H. M. (1964) The theory and operation of settling tanks, *Trans. Inst. Chem. Eng.* **42**, T 158.
40. STAIRMAND, C. J. (1949) *Engineering*, London, **169**, 409.
41. SHEPHERD, C. B. (1940) *Ind. Eng. Chem.* **32**, 1246.
42. DAVIDSON-HARRISON (1963) *Fluidised Particles*, Cambridge University Press.
43. *Aspects of Fluidisation* (1959) Abstracts of papers presented at Manchester Symposium, *B. Chem. Eng.* **4**, 289.
44. LAPPLE, C. E. (1940) *Ind. Eng. Chem.* **32**, 605.
45. MICHELL, S. J. (1965) *Particle Dynamics Processes*, Symposium Preprints, Huddersfield College of Technology.
46. CREMER, H. W. and DAVIES T. (1957) *Chemical Engineering Practice*, Butterworths Scientific Publications, London.

THE SI UNITS

Some derived SI units

Physical quantity	SI unit	Symbol
Force	Newton	$N = kg\ m/s^2$
Energy (work, heat)	Joule	$J = N\ m$
Power	Watt	$W = J/s$

Some prefixes in use with the SI units

10^{-6}	micro	μ
10^{-3}	milli	m
10^3	kilo	k
10^6	mega	M

Some conversion factors
(An asterisk (*) denotes an exact relationship)

Length

* 1 in.	25·4 mm
* 1 ft	0·3048 m
* 1 yd	0·9144 m
1 mile	1·6093 km

Area

* 1 in²	645·16 mm²
1 ft²	0·0929 m²
1 yd²	0·8361 m²
1 mile²	2·590 km²
1 acre	4046·9 m²

Volume

1 in³	16,387·0 mm³
1 ft³	0·02832 m³
1 U.K. gal	0·004546 m³
1 U.S. gal	0·003785 m³

Mass

* 1 kg	10^3 gm
1 oz	28·352 gm
1 lb	453·592 gm
1 cwt	50·802 kg
1 ton (U.K.)	1016·06 kg

Density

1 lb/ft^3	16·019 kg/m^3
1 lb/U.K. gal	99·776 kg/m^3
1 lb/U.S. gal	119·83 kg/m^3

Force

* 1 dyne	10^{-5} N
1 poundal	0·1383 N
1 lb-f	4·4482 N
1 kg-f	9·8067 N
1 ton-f	9·9640 kN

Viscosity, dynamic

* 1 poise (1 gm/cm sec, 1 dyne sec/cm^2)	0·1 N sec/m^2
1 lb/ft sec (1 poundal sec/ft^2)	1·4882 N sec/m^2
1 lb/ft hr (1 poundal hr/ft^2)	0·4134 mN sec/m^2

Viscosity, kinematic

* 1 stokes (1 cm^2/sec)	10^{-4} m^2/sec
1 ft^2/hr	25·806 mm^2/sec

Pressure

* 1 bar (10^6 dynes/cm^2)	10^5 N/m^2
* 1 at (1 kg-f/cm^2)	98·0665 kN/m^2
* 1 at (standard)	101·325 kN/m^2
1 psi (1 lb-f/in^2)	6·8948 kN/m^2
1 psf (1 lb-f/ft^2)	47·880 N/m^2
1 ton-f/in^2	15·444 MN/m^2
1 torr (1 mm Hg)	133·32 N/m^2
1 in Hg	3·3864 kN/m^2
1 in water	249·09 N/m^2
1 ft water	2·989 kN/m^2

Energy (work, heat)

* 1 erg	10^{-7} J
1 ft poundal	0·04214 J
1 ft lb-f	1·3558 J
* 1 cal (international table)	4·1868 J
1 Btu	1055·06 J
1 hph	2·6845 MJ
* 1 kWh	3·6 MJ

Power

* 1 erg/sec	10^{-7} W
1 hp (British)	745·70 W
1 hp (metric)	735·50 W
1 ft lb-f/sec	1·3558 W
1 Btu/hr	0·2931 W

Surface tension (energy)

* 1 dyne/cm (1 erg/cm²)	10^{-3} J/m²

Moment of inertia

1 lb ft²	0·04214 kg m²

Momentum

1 lb ft/sec	0·1383 kg m/sec

Angular momentum

1 lb ft²/sec	0·04214 kg m²/sec

Specific heat (heat capacity)

* 1 Btu/lb °F (1 Chu/lb °C, 1 cal/g °C)	4·1868 kJ/kg °K

Heat transfer coefficient

1 Btu/h ft² °F	5·6783 W/m² °K

Thermal conductivity

1 Btu/h ft °F	1·7307 W/m °K

Approximate values of some properties of water (at 18°C), *and air* (at STP)

	Water	Air
Density (kg/m³)	10^3	1·3
Viscosity (N sec/m²)	10^{-3}	$1·7 \times 10^{-5}$
Specific heat (kJ/kg °K)	4	1
Thermal conductivity (W/m °K)	0·6	0·024

INDEX

Page references in **bold** type are to pages on which an item of more importance occurs.

$$\frac{\text{m cm}}{\text{s}}$$

byosoku

5 m/s

byosoku 5 metre

毎時 5 kg m

5 km/h

$$\frac{m^2}{s} =$$